SPACE FLIGHT:
CROSSING
THE
LAST FRONTIER

A Mercury capsule is installed atop a Redstone rocket in 1960. Project Mercury was the first American effort to launch humans into space.

SPACE FLIGHT: CROSSING THE LAST FRONTIER

Jason Richie

The Oliver Press, Inc.
Minneapolis

I wish to extend a heartfelt thank-you to Christopher Rink, Langley Research Center, Office of Public Affairs, for his help in setting up interviews with John Houbolt and Max Faget.

—*Jason Richie*

The Oliver Press, Inc.
Charlotte Square
5707 West 36th Street
Minneapolis, MN 55416-2510

Library of Congress Cataloging-in-Publication Data
Richie, Jason, 1966-
Space flight : crossing the last frontier / Jason Richie.
p. cm. — (Innovators ; 10)
Includes bibliographical references and index.
    Summary: Profiles seven engineers and scientists who made space flight possible, including Robert Goddard, Wernher von Braun, Max Faget, John Houbolt, William Pickering, Sergei Korolev, and Dale Reed.
ISBN 1-881508-77-3 (library binding)
1. Aerospace engineers—Biography—Juvenile literature. 2. Space flight—History—Juvenile literature. 3. Manned space flight—History—Juvenile literature. [1. Aerospace engineers. 2. Space flight—History.] I. Title. II. Series.
TL788.5 .R55 2002
629.4'09'2—dc21
[B]
                                        2001036507

ISBN 1-881508-77-3
Printed in the United States of America
08 07 06 05 04 03 02   8 7 6 5 4 3 2 1

CONTENTS

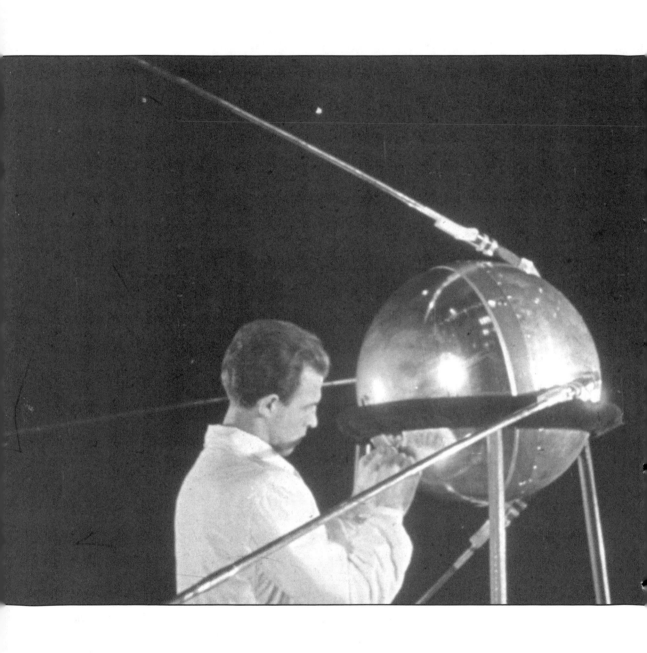

"All Right. Let's Get On With It!"

History records October 4, 1957, as the day *Sputnik*, the first artificial satellite, reached orbit. That's true, of course, and it was a spectacular achievement for the Soviet Union. But history could have played out differently. More than a year earlier, on September 20, 1956, an American rocket stood ready on a launch pad at Cape Canaveral, Florida. A collaborative effort between Wernher von Braun's rocket team at Marshall Space Flight Center in Alabama and William Pickering's Jet Propulsion Laboratory in California, the multistage Jupiter-C had the power to thrust its nose cone (the top section of the rocket) into orbit—except for one deficiency. Its final stage, or section, was inoperative. When the rocket flew that day, its nose cone reached nearly 700 miles into space—much farther than necessary to orbit—but only at a speed of about 16,000 miles per hour. That was some 1,500 mph shy of orbital velocity, a boost the fourth stage would have added if it had been

All right. Let's get on with it!
—Thomas Keith Glennan, first administrator of NASA, upon the launch of Project Mercury in October 1958

A Soviet engineer working on the first artificial satellite to circle Earth. Although only 184 pounds and 23 inches in diameter, Sputnik 1 was visible in the night sky and transmitted a beeping signal for 21 days before it reentered the atmosphere and burned up.

satellite: a small celestial body that orbits a larger one, such as the Moon traveling around Earth. An artificial satellite is a human-made object, such as a weather satellite, that orbits Earth.

orbit: the path of one body as it circles another, such as Earth around the Sun, or a satellite around Earth

space: the vast region beyond the atmosphere of Earth

astronomy: the science that studies the stars, planets, and all other celestial bodies

working. The nose cone plunged into the Atlantic Ocean after a long, looping flight of 3,300 miles.

Hardware malfunctions were common enough in the early days of the space program. On this day, however, the big rocket did not misfire. Its fourth stage had been deliberately disabled and filled with sand. An act of sabotage by Soviet agents? No. The U.S. Army general in charge of the launch, acting on the wishes of the president himself, had ordered the modification.

Von Braun and Pickering knew beforehand that their missile would not be allowed to reach orbit; President Dwight Eisenhower believed that a military satellite would provoke the Soviets. Politicians and generals were calling the shots and, as usual, their agenda had little to do with advancing our knowledge of space.

No one knows when the spark to study the stars first ignited. The urge to understand the cosmos is likely as old as human awareness itself. Astronomers from the earliest known civilizations, ancient Egypt and Sumer, accurately mapped the stars more than 4,500 years ago. Succeeding civilizations—in Mexico in North America, in Peru in South America, in Cambodia in Southeast Asia, and on Easter Island in the Pacific Ocean—showed the same obsession with the night sky, building temples and whole cities or drawing immense images in the ground to represent what they saw there.

Many principles of modern astronomy date to ancient Greece. Haraclides (388-315 B.C.), for instance, found that at least some of the known planets revolve around the Sun and not Earth. He also attributed the nightly trek of the stars across the sky to a spinning motion of Earth on its axis. Aristarchus (310-230 B.C.) was first to suggest that Earth was just another planet moving around the Sun. The idea proved too radical to gain widespread acceptance at the time—or for another 1,800 years!

Polish astronomer Nicolaus Copernicus (1473-1543) resurrected Aristarchus's Sun-centered idea to describe the motions of the planets. Other "Copernicans" followed: German Johannes Kepler (1571-1630), who found that planetary orbits were not perfect circles but rather oval-shaped ellipses; Galileo Galilei (1564-1642) of Italy, whose invention of the "optik tube" (a telescope) in 1609 changed stargazing forever; and Isaac Newton (1642-1727), who expanded upon Kepler's work to formulate three laws of universal motion.

Newton's laws, published in 1687, explained for the first time why stars, planets, and moons move as they do. Newton wrote that all bodies remain at rest or moving in a straight line at a constant speed unless acted upon by some force. This was his first law, the law of inertia. The Moon, for instance, would fly off in a straight line if not for Earth's gravity trapping it in orbit. At the same time, the force of gravity isn't strong enough to pull the Moon all the way to Earth. The Moon's inertia—its inclination to remain moving away—balances against Earth's attraction. The same principle applies to every body in the universe.

Copernicus revolutionized science when he realized that the motion of the planets could only be explained if the Sun was at the center of the solar system.

The famous falling apple that inspired Newton did not hit his head, but it caused him to ask a question: If an apple falls to the ground, why doesn't the Moon?

Born in France, Jules Verne first studied law, then wrote texts for operas before popularizing a new type of literature: science fiction.

The rocket above, an illustration in one of Verne's books, looks very much like a modern rocket.

The law of inertia also explains how a satellite might be sent into orbit. Newton drew a simple diagram to illustrate his point. In the sketch, a cannon fires a ball from atop a tall hill. The ball travels a short distance and hits the ground. In successive shots, the cannon fires with greater force and the balls fly farther and farther each time. Finally, the cannon applies enough force that the ball flies so high that it does not fall back to Earth. But at the same time, the ball does not have enough energy to fly deep into space. The cannonball's inertia balances against Earth's gravity, and the cannonball achieves orbit.

Newton used the cannon diagram as a conceptual model only. No real gun, then or now, could launch its shot fast enough to put it in orbit. So, how could an object be made to break free of its earthly chains? Part of the answer can be found in Newton's laws, but it wasn't spotted at the time or for the next two centuries.

Science fiction writer Jules Verne (1828-1905) moved towards the truth in 1870 with *Around the Moon*. True, his fictional heroes relied on a cannon shot to get to the Moon, but their manned shell used small rockets to maneuver through space. Verne's attempts at realism were "fictionalized science lessons," as one historian called them. His tales fired the imaginations of readers around the world, including Konstantin Eduardovich Tsiolkovsky, born in Izhevskoye, Russia, in September 1857. Poor as a child, Tsiolkovsky educated himself in mathematics and science. By 1876, he had learned enough to become a schoolteacher. In his spare time, he scribbled thoughts of space travel and its effects on humans and machines in a small notebook.

Tsiolkovsky was the first true rocket scientist. He calculated the speeds necessary to reach orbit and to escape Earth's gravity. Once he knew the numbers, he searched for a way to achieve them. He found it in Newton's third law of motion: For every action there is an equal and opposite reaction. "Consider a cask filled with a highly compressed gas," Tsiolkovsky wrote in 1883. "If we open one of its taps the gas will escape through it in a continuous flow; the elasticity of the gas pushing its particles into space will continuously push the cask itself."

"I have worked out several aspects of the problem of ascending into space by means of a reactive device," wrote Konstantin Tsiolkovsky. *"My mathematical conclusions . . . show that with such devices it is possible to ascend into the expanse of the heavens, and perhaps to found a settlement beyond the limits of the earth's atmosphere."*

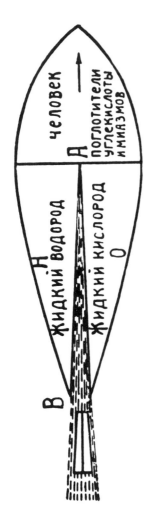

Tsiolkovsky sketched this rocket in 1903. Labeled in Russian, his design featured a nozzle at one end (B) and room for a pilot and a device to clean the air in the top section (A). He proposed liquid hydrogen and liquid oxygen for fuel (rear sections, H and O).

Did Tsiolkovsky really think a gas-filled barrel could launch satellites and people to space? Yes, but not just any barrel. Tsiolkovsky had in mind one made of metal, pointed at one end and with its "taps"—or nozzles—at the other. Inside the barrel, the expanding gas, the product of burning fuel, would push out in all directions. Since the gas would be allowed to escape in only one direction, the barrel would have no choice but to move in the opposite direction. Tsiolkovsky's barrel, of course, was a rocket.

Konstantin Tsiolkovsky knew rockets could operate in space. Rockets didn't need air to push against as wings and balloons did, and they could carry with them the oxygen needed to burn fuel. Unfortunately, the rockets of his day could never reach space. They were fueled with gunpowder and burned too inefficiently. Tsiolkovsky's solution was to power his theoretical rockets with liquids—specifically, liquid hydrogen and liquid oxygen, which he correctly calculated to be the most efficient and powerful propellants available. It was a brilliant idea and one that could power space flight.

But in Newton's third law and his own liquid-fuel rocket, Tsiolkovsky had found only part of the answer. Conquering space takes more than brilliant ideas; it takes money and lots of it. The only repositories of enough cash were national treasuries. Not until the Communists took over Russia in 1917 and offered Tsiolkovsky and his followers government support did any practical advances occur. Tsiolkovsky was content with theorizing about human space flight until his death in 1935, but other Soviet rocketeers who followed, like Sergei Korolev, began to actually build what Tsiolkovsky envisioned. Korolev helped design the Soviet Union's first liquid-fuel rocket, which flew in 1933. In the 1950s, he oversaw development of the world's first intercontinental ballistic missile (ICBM), the R-7. He used it in 1957 to launch another of his creations, *Sputnik*, the first artificial satellite.

Korolev's 1933 rocket was not the world's first. That was launched by American Robert Goddard seven years earlier. Goddard was every bit an American Tsiolkovsky and more. Independently of the Russian, he discovered the same secrets of space flight. Goddard came to realize something else Tsiolkovsky had learned, that building rockets without a lot of money is next to impossible. He labored in near obscurity until his death in 1945, his tremendous talent untapped.

Wernher von Braun of Germany never faced the same problem. As a teenager in 1925, von Braun became addicted to the idea of space flight after reading *The Rocket into Planetary Space*, by German scientist Hermann Oberth. Like Tsiolkovsky and Goddard, Oberth had worked out the problems of space flight himself. Von Braun became Oberth's prize pupil and later, as a 20-year-old in 1932, was recruited by the German army to oversee development of a long-range ballistic missile. The Germans gave von Braun what he needed. The result was the V-2, the world's first ballistic missile, initially launched in 1942 to a low-space altitude of over 50 miles.

Robert Goddard and his first rocket, which flew in 1926. Weighing only 16 pounds, more than 10 of which were fuel, the rocket launched a new era of human achievement.

altitude: the height of something above a specific reference level

After von Braun's creation, the pace of innovation quickened dramatically. The enabling force, however, was not humanity's desire to know the stars. Luckily for rocketeers like von Braun, who continued his work in the United States after Germany's defeat in World War II, America became engaged in a non-shooting "Cold War" with the Soviet Union. Space became another theater in the conflict, "high ground" both sides desired to capture.

Why was this? It was all about the U.S. and the Soviet Union gaining respect—from the rest of the world and from each other. Both hoped to win the allegiance of nations not aligned with either. Each hoped to deter an attack from the other by building nuclear weapons. To create and deliver an atomic or hydrogen bomb takes a high level of technological advancement. Space provided an excellent arena in which each power could show it had attained the necessary level to destroy the other. That is why the launch of *Sputnik* was so much of a shock to Americans.

Respect is also why the United States could not let *Sputnik* go unanswered. America's first response was to shoot its own satellite into orbit. Designed by the California Institute of Technology's Jet Propulsion Laboratory, led by William Pickering, *Explorer 1* was launched in January 1958. Pickering's lab followed that with America's first Moon probes, then the world's first successful robot explorers, sent to Mars and Venus. Meanwhile, the National Aeronautics and Space Administration (NASA) was directed to send the first human into space. Max

probe: a spacecraft designed to leave Earth and explore and photograph planets in our solar system, their moons, and other objects in space

Faget and his engineering team designed America's first spacecraft, the Mercury capsule, which took astronaut Alan B. Shepherd into space on May 5, 1961. But Faget's effort didn't achieve the desired end, because Korolev had beaten America to the punch by launching the first manned spacecraft three weeks earlier.

So the United States upped the ante by committing to send a piloted expedition to the Moon by

Beaming with success, William Pickering, James Van Allen, and Wernher von Braun (left to right) held a full-sized model of Explorer 1 when they announced that the satellite was in orbit.

The three Apollo 11 *astronauts began their historic journey the morning of July 16, 1969, lifting off from launch pad 39A at Kennedy Space Center in Florida.*

the end of the decade. Again Faget oversaw construction of the U.S. spacecraft. Designing the Moon rocket, the massive Saturn 5, was Wernher von Braun. This time the Americans would not be denied; only they could afford the tremendous costs involved. Employing a travel technique called lunar orbit rendezvous championed by NASA engineer John Houbolt, the astronauts of *Apollo 11* journeyed to the Moon and back in the summer of 1969.

Even as America was taking great strides in space, it was losing ground in yet another theater of the Cold War, Southeast Asia. The communists in North Vietnam were waging war against South Vietnam. The U.S. failed to stop the North Vietnamese and finally withdrew its troops in 1973. Americans had turned against their nation's involvement in the struggle long before then, however. Their general discontent reached even the space program. They simply tired of costly contests. The last three Apollo missions, 18 though 20, were scrapped, and soon thereafter any plans for a piloted mission to Mars.

A permanent space station was scuttled, too, but not the shuttle intended to support it. Max Faget led the design of that, too. It flew for the first time in 1981. As the Faget team members created the shuttle, they used data gleaned from a series of tests carried out in the early 1960s on weird-looking aircraft called "lifting bodies." Originally conceived in the 1950s as a possible space capsule, the first lifting body was potato-shaped and flew without wings. The idea didn't originate with NASA engineer Dale

Reed, but he rescued it from the junk heap and fought to build it.

That's not the end of Reed's story. In 1996, NASA announced that a lifting-body design had won the competition to replace the aging shuttle. Furthermore, a small lifting-body "lifeboat" would be built for use on the international space station, *Alpha*. But, amid spiraling development costs and budget cuts in 2001, NASA canceled the lifting-body shuttle. And what of Reed? He invented an airplane capable of flying through the atmosphere of Mars. If men and women from Earth do land there someday, they just might take off again to explore the red planet in a Dale Reed-inspired Marsplane.

But that's a big "if." There currently seems to be little desire among political leaders to fund big-budget space travel. The end of the Cold War reduced the military value and prestige factor associated with sending people into space.

As for exploring for the sake of knowledge, that has rarely motivated national leaders. But although generals and politicians may not want "to go to Saturn to count rings," as William E. Burrows has argued, many scientists and explorers long to learn more about our solar system and beyond. Like the innovators featured in this book, these dreamers will never stop searching for the ways and means to venture across the last frontier.

Rocketeers . . . could dream about space travel all they liked, but they were shackled to a common and unavoidable dilemma: Their lifeblood, their financial sustenance, flowed from politicians and generals. . . . No politician or general wanted to go to Saturn to count rings.
—William E. Burrows, space historian

CHAPTER ONE

Robert Goddard and the Liquid-Fuel Rocket

For Robert Goddard, the fixation on space flight began early. In 1898, the 16-year-old read *The War of the Worlds*, a science-fiction novel by H. G. Wells. As Goddard recalled, the fantastic tale of Martians traveling millions of miles through space to attack Earth "gripped my imagination tremendously." Within a year he devoted his life to inventing a device that could navigate the same interplanetary route.

Goddard succeeded brilliantly. He was the first to build and launch a liquid-fuel rocket, refined versions of which have pushed astronauts to the Moon and robot probes to the very limits of our solar system. These are facts, available in any thorough history on the subject. They're usually preceded by labels describing Goddard as the "father" of American rocketry, even of all modern rockets. Such titles have been used so often and for so long that they are now regarded as truth. Unfortunately, they aren't. To understand why is to

interplanetary space: space among the planets, within the solar system and outside the atmosphere of any planet or the Sun

Robert Hutchings Goddard (1882-1945) works on an early form of rocket with steel chamber and nozzle in 1915. Eleven years later, he launched the first liquid-fuel rocket towards space.

19

British writer H. G. (Herbert George) Wells (1866-1946) influenced many people of his time with his progressive social ideas and scientific fantasies.

"My father and I were great pals," Goddard later wrote, and the two often canvassed the countryside in search of frogs, fish, and other wildlife to observe.

—Milton Lehman, author of *This High Man: The Life of Robert H. Goddard*

know both a great innovator and the real story behind the birth of the Space Age.

Robert Hutchings Goddard was born in Worcester, Massachusetts, on October 5, 1882. Early on, the boy showed his inventive streak. When just five, he jumped repeatedly from atop a fence, holding part of an old battery in one hand. Before every attempt he "charged" the battery by scuffing his feet on a piece of carpet and producing sparks. He was sure he flew farther each time. Nahum and Fannie Goddard encouraged their son's curiosity. They bought Robert a telescope, a microscope, a subscription to a science magazine—most anything to inspire his experiments.

Just weeks into his seventeenth year in October 1899, Goddard had a vision of space flight that set him upon his historic trajectory. Perched in a cherry tree, intent on pruning dead limbs, Goddard imagined a mechanical device that could soar to Mars and saw himself launching a small model from the meadow outstretched before him. "I was a different boy when I descended," he later wrote, "for existence at last seemed very purposive." He marked the day, October 19, in his diary every year, calling it "Anniversary Day." Despite his strong motivation, Robert struggled to build his space machine. His limited knowledge of physics, mostly the result of frail health that often prevented him from attending school, was holding him back.

Redoubling efforts to educate himself, Robert chanced upon the writings of Isaac Newton (1642-1727), the English physicist and mathematician

Although Robert Goddard is not sitting in the same tree where he had his vision, this photograph with his mother, Fannie Louise Goddard, was taken shortly after the event.

whose monumental laws of motion (1687) explain why objects on Earth and in space move as they do. Briefly, the three laws are: 1) an object remains at rest or moving in a straight line at a constant speed unless acted upon by a force; 2) the change in motion is proportional to the force involved and will be in the same direction as the force; and 3) for every action there is an equal and opposite reaction. Careful study of the laws convinced Goddard he had found the keys to space travel.

All three laws are important to space flight, but law three especially concerned Robert as he searched for a device that could propel itself through space. To understand why, consider a firearm. When a gun is fired, gas builds up inside before finally exploding from the open end of the barrel (pushing the bullet out as it goes). This explosion is the action and it causes a reaction: the gun "kicks" in the opposite direction, against the shooter's hand or shoulder. In 1904, his senior year at Worcester's South High School, Robert proposed his first spacecraft—a cannon that was propelled up by firing its explosive blast earthward.

Goddard considered the idea over the next four years while earning a bachelor's degree in science at Worcester Polytechnic Institute. In 1909, as a second-year graduate student at Clark University in Worcester, he abandoned cannons for something more practical—rockets. Like cannons and other firearms, rockets are reactive devices. They move in one direction as exhaust gases from their burning fuel explode in the opposite direction. Like firearms, rockets were fueled by gunpowder in Goddard's day.

oxidizer: an oxygen-rich substance that enables fuels to burn in space

Gunpowder can burn in space because one of its ingredients, potassium nitrate, is an oxidizer. (Oxygen is required for nearly all combustion. Jet and internal combustion engines use oxygen in the air to burn their fuel. Rockets must carry their own source of oxygen.) Unfortunately, gunpowder rockets weren't powerful enough to reach space. Robert Goddard theorized about new types of rockets, including solar and atomic, but little was known about capturing and using such exotic energy sources. He also considered liquid propellants, which promised greater performance than powders.

Goddard received a doctorate in physics from Clark University in 1911. Following two more years of study on a research fellowship at Princeton, he joined Clark's faculty in 1914.

The same year Goddard began teaching physics at Clark, the U.S. government issued the professor two patents for his inventions that firmly established him as America's leading rocket pioneer. The patents covered feed systems, which route liquid fuels from tanks into the combustion chambers, where the fuels ignite and burn, and nozzles, designed to vent the exploding exhaust gases rearward to cause forward motion.

Also covered was the concept of the multistage rocket. Goddard understood that a rocket would need an enormous amount of fuel and power to lift itself and a payload, or cargo, into space. As the rocket burned the fuel, more of its body would become dead weight. If the useless mass could be discarded in stages, there would be less rocket to lift. The same payload could then be launched with a smaller rocket. All modern rockets use feed systems, combustion chambers, nozzles, and multiple stages.

Goddard, hoping to gain additional funding for research, submitted a report of his work in 1916 to the Smithsonian Institution in Washington, D.C. In the account, Goddard described how to build a

propellant: the combination of fuel and oxidizer that a rocket engine burns to achieve thrust

As far as Goddard knew, no one else had ever considered a liquid-fuel rocket—or solar or atomic rockets, for that matter. He had never heard of Konstantin Tsiolkovsky.

gravity: the attraction of the mass of a celestial body, such as Earth, for other bodies. Gravity creates a sense of weight and causes objects to fall.

two-stage powder-fuel rocket for atmospheric research and how such a rocket could reach "infinite altitude"—that is, break free of Earth's gravity. To prove the feat, Goddard suggested crashing the rocket on the Moon and observing the flash through powerful telescopes.

Much impressed, the Smithsonian granted Goddard the modest sum of $5,000. Barely had the professor drawn on his account when America entered World War I in April 1917. From then until the armistice of November 1918, Goddard devised rocket weapons for the U.S. Army. The generals especially loved his shoulder-fired rocket. Their interest fizzled with the end of the war, however, and it would take another world war 20 years later before Goddard's weapon—which soldiers called a "bazooka" for the strange sound it made when fired—finally caught on.

In 1920, the Smithsonian published Goddard's 1916 grant proposal—"A Method of Reaching Extreme Altitudes." Within weeks, newspapers across the country were hyping—and sometimes ridiculing—his conclusion that rockets could fly to the Moon. "Modern Jules Verne [a popular science fiction writer of the previous century] Invents Rocket to Reach Moon," ran one sensational headline. An editorial in *The New York Times* even suggested Goddard lacked the basic knowledge of physics "ladled out daily in high schools."

Goddard was glad space flight sparked the public's interest, but the whole affair greatly embarrassed

him. Shy and secretive to begin with, Goddard reacted to the uproar by further isolating himself.

Ironically, Goddard's rockets weren't reaching anywhere near "extreme altitudes." The inventor simply could not make powder burn evenly and predictably. In January 1921, he switched to liquids. It was humanity's first practical step toward space.

Professor Goddard explains his idea of sending a rocket to the Moon.

THE BREAKTHROUGH

liquid-fuel rocket: a rocket propelled by liquid chemical fuels and oxidizers

liquid hydrogen: the liquid form of hydrogen created by cooling the gas to extremely low temperatures. Hydrogen is the lightest, simplest element and burns easily.

liquid oxygen: the liquid form of oxygen created by cooling the gas to extremely low temperatures. Oxygen is one of the most abundant elements on Earth.

Together, liquid hydrogen and liquid oxygen yield the highest output of energy of any fuel-oxidizer combination. Liquid hydrogen has the greatest heat content of any known fuel.

thrust: the driving force that pushes a rocket engine forward

The pioneer's first order of business was to pick and test liquid propellants. Goddard needed both fuel and oxidizer, an oxygen-rich substance that enables fuels to burn in space. In theory, liquefied hydrogen was the best fuel, but it was a relatively recent innovation and still very scarce in the 1920s. He instead used easily obtainable hydrogen compounds, like ether and gasoline. For an oxidizer, the professor chose liquid oxygen. Although the pale blue liquid was available, it was expensive and dangerous; while not flammable itself, anything it splashed on could become a firebomb. "Goddard's first test was a tremendously courageous thing," recalled a later rocket pioneer, G. Edward Pendray.

Early tests showed that the fuel and oxidizer burned smoothly and controllably, not explosively like powders. As the tests progressed over several months, the professor and his assistant injected the mixture into a small combustion chamber, where it ignited and spewed hot exhaust from the attached nozzle. When the primitive engine generated thrust in the opposite direction, Goddard knew his liquid-fuel rocket could work.

Next up for the professor was a problem that would plague him for years: constructing a pump that was powerful and sturdy enough to force super-chilled liquid oxygen into the combustion chamber. Liquid oxygen ranges in temperature from -297° F to -361° F. At those temperatures, metals become very brittle. Goddard's journal is replete with entries like, "Got pump outfit ready. . . . It jammed;" or, "Tried pump. . . . It bound." He tried at least five different types of pumps and failed each time.

Then there was the problem of too much heat in the combustion chambers, which caused them to melt. In early 1923, Goddard developed a process called "regenerative" cooling. He wrapped a jacket of metal around the chamber, then routed the super-cold liquid oxygen through the jacket on its way to the chamber. This cooled the chamber. Goddard invented "curtain" cooling around the same time. He ringed the inside top of the chamber with tiny holes, through which he pumped fuel or liquid oxy-gen. The liquid coated the chamber wall to form a protective curtain against the fiery combustion. Both regenerative and curtain cooling would be used by later rocketeers.

As Goddard doused one fire after another, he did so under the constant threat of bankruptcy. His modest annual salary of $2,000 carried him only so far, and the Smithsonian's grant of 1916 and several smaller endowments from Clark University were running out. Smithsonian officials gave him an additional $2,500 in 1924, but they demanded a quick return on their investment. It is the "same old story," Goddard wrote a friend, "of no support until results are had, and no results unless sufficient sup-port is had."

Considerations of time and money forced Goddard to settle for less than his ideal machine, at least for the first flight. The rocket he built over the winter of 1925-1926 did not carry a pump to drive

Sir James Dewar (1842-1923), British chemist and physicist, was the first person to produce hydrogen in liquid form in the late 1880s. He also invented the Dewar flask—the first vacuum, or thermos, bottle—as a container for his liquefied gases.

Both oxygen and hydrogen are used as liquids instead of gases because liquids are denser, and therefore more can fit into a fuel tank. The gases must be cooled to extremely low temperatures to transform them into liquid form.

the liquid propellants, but rather an extra tank of gaseous oxygen to pressure-feed the fuel and oxidizer into the combustion chamber. There was no curtain or regenerative cooling on the first rocket, either. Goddard coated the combustion chamber and nozzle with heat-resistant ceramic, a quick fix that would never hold during long flights. Finally, the rocket had no instruments for guidance and control.

Still, what Goddard and his assistant, Henry Sachs, assembled and fueled on the crisp, wintry morning of March 16, 1926, was state-of-the-art technology. The rocket weighed ten and a quarter pounds fully fueled and stood ten feet tall. At its base sat the fuel and oxidizer tanks, situated one atop the other. Two tubes carrying the gasoline and liquid oxygen ran upward from these to the rocket's engine, which occupied the top two feet or so. (Goddard hoped the engine-on-top arrangement would lend stability in flight. It didn't. Subsequent Goddard rockets would have their engines in the tail.) Finally, holding the fragile-looking contraption upright was a tubular launch tower, actually a converted mail-order windmill.

Preparation ended around 2:30 that afternoon. While Sachs lit a small alcohol burner beneath the rocket's engine with a blowtorch, Esther Goddard (Robert's wife since 1924) activated a small motion-picture camera. The professor pumped gaseous oxygen into the fuel and oxidizer tanks, which in turn pushed the liquids into the engine. They ignited and burned, slowly building up the power, or thrust, necessary to lift the rocket. After about 90 seconds, Goddard released the rocket's restraint and ran for cover. "The rocket did not rise at first," he wrote the next day, "but the flame came out, and there was a steady roar." Suddenly, "at express-train speed," it shot free of the tower, arched over, and struck the snow some 180 feet away. The whole trip took two and a half seconds.

Henry Sachs ignites Goddard's rocket on March 16, 1926. "It was the most beautiful sight in the world, seeing the rocket take off," Esther Kisk Goddard wrote. "We slogged jubilantly through the mud of Aunt Effie's cabbage patch toward the wreckage."

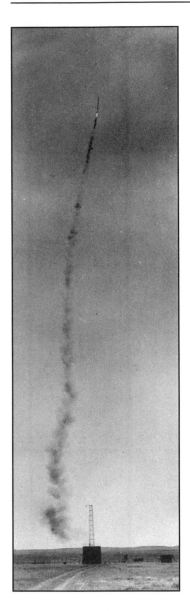

One of Goddard's A-series rockets pierces the New Mexican sky in 1935.

THE RESULT

Goddard's second flight came two weeks later, on April 3, 1926. The fourth flight, in July 1929, was the first to carry instruments: a barometer to measure atmospheric pressure; a thermometer to gauge temperature (which, taken along with pressure, gives altitude); and a camera to record these readings. The locals were not impressed. They thought the rocket's fiery crash was an airplane. A new round of sensational headlines—and banishment by the fire marshal—prompted Goddard to move.

The professor headed west to Roswell, New Mexico. The desert around Roswell was sparsely populated and relatively flat, with moderate temperatures and clear skies year round—in short, ideal conditions for testing rockets. Goddard was originally bankrolled in this renewed effort by a two-year, $50,000 grant from the wealthy Guggenheim family. As the Great Depression worsened in the early 1930s, the Guggenheims curtailed their support, even briefly terminating funding in July 1932. The yearly stipends, now reduced to $18,000, resumed in August 1934. With this, Goddard had to support himself and Esther, rent a ranch, pay four assistants, and purchase all necessary supplies and tools needed to build his rockets.

Over the next several years, Robert Goddard's achievements were many and unprecedented. A self-stabilizing rocket flew successfully for the first time in March 1935, with an onboard gyroscope activating tail-mounted steering vanes. In July 1938, the first rocket with "gimbal" steering (no vanes, just a movable engine) lifted off. In August 1940, Goddard finally launched a rocket with its fuel and oxidizer fed by a pump—yet another innovation that would show up in later liquid-fuel rockets.

After the United States entered World War II in December 1941, Goddard worked for the military, mostly on JATO (Jet Assisted Takeoff) rockets designed to boost airplanes during takeoff. The frustrated pioneer thought the military had "no particular interest in rockets or rocket propulsion."

The great inventor did not survive the war's end. In May 1945, doctors diagnosed throat cancer. Goddard died of the affliction on August 10, 1945, four days before Japan's surrender ended the war.

Famed transatlantic flier Charles Lindbergh, enthralled with Goddard's work, secured the Guggenheim grant through his friendship with Harry, son of philanthropist Daniel. Harry Guggenheim (second from left) and Lindbergh (second from right) visited Goddard (center) in New Mexico in 1935. Also pictured are two of Goddard's assistants, A. W. Kisk (left) and N. T. Ljungquist (right). Kisk was Esther Goddard's brother.

G. Edward Pendray works on a rocket in about 1932. This rocket was never fired.

Robert Goddard's reputation as a brilliant innovator is well deserved. By the time of his death, he had been awarded 48 patents involving all aspects of rocketry. An additional 166 were granted after his death, much to the credit of Esther, who transcribed and organized her husband's papers for the couple's patent attorney.

Many historians have made the case, as one American space official has, that "every liquid-fuel rocket that flies is a Goddard rocket." If by this they mean that every rocket owes its existence to Goddard, they are wrong. Although all liquid-fuel rockets use many of the same design principles as Goddard's rockets, their designers came by these principles independently of the secretive Goddard.

These include not only European innovators like Hermann Oberth, Wernher von Braun, and Sergei Korolev, but also rocket enthusiasts from the United States. One of the Americans was G. Edward Pendray, who was vice president of the American Interplanetary Society (AIS) when he toured Germany in 1931. Goddard had shunned the AIS, so Pendray learned what he could from the Germans. It was German technology, not Goddard's, that contributed to the launching of an AIS liquid-fuel rocket in May 1933—the first American rocket outside of Goddard's. Another AIS member, James Wyld, read enough German to learn of a cooling technique that substantially improved the operation of the society's engines. Wyld went on to found a company that built the liquid-fuel rocket engine that powered the Bell X-1, the aircraft that first broke the sound barrier in 1947.

Yet another American pioneer was Frank Malina, a student at the California Institute of Technology (Caltech) in the 1930s. Unable to learn much from Goddard, Malina, too, scoured the works of the Europeans. That research, plus his own creativeness and experiments with fellow enthusiasts from Caltech, gave Malina the confidence and knowledge he needed to take the lead in founding Caltech's Jet Propulsion Laboratory (JPL) in 1942. As we shall see, JPL became the leader in the development of American rockets.

Besides his own character, Goddard fell victim to weak government demand for his product. Because space flight was still considered science fiction, building rockets for that purpose was never considered. During most of Goddard's lifetime, the U.S. was uninvolved in world affairs and had little need for long-range weapons. Through most of his career, Goddard relied on modest grants. While usually steady, his funding was not substantial enough to allow Goddard and his rockets to reach their potential.

Contrast this to the Germans, who realized the destructive potential of rockets as early as the 1920s. Lavish government backing enabled Wernher von Braun to launch the first ballistic missile in October 1942. On that day, the German V-2 soared to an altitude of about 53 miles, the edge of space. Goddard's best effort, in 1935, reached less than a mile and a half. Had Goddard had similar support, the Americans might have reached space first. The Space Age to follow would have been his offspring, not the Germans'.

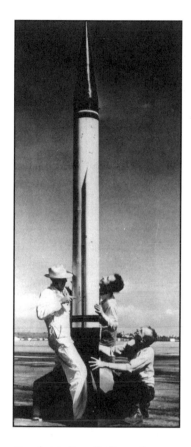

Frank Malina (left) makes adjustments on a WAC Corporal rocket at White Sands, New Mexico, in the 1940s.

CHAPTER TWO

Wernher von Braun and the Moon Rocket

While Robert Goddard was developing rockets in the United States in the 1920s, parallel efforts sprouted in Europe. There, pioneers behind the upsurge gained inspiration not from the secretive American, but from outgoing German physicist Hermann Oberth. *The Rocket into Planetary Space*, a 92-page book written by Oberth and published in 1923, was densely packed with calculations that mathematically proved liquid-fuel rockets could travel through airless space and reach other planets. Better yet, Oberth figured people could survive the trip, too. *The Rocket into Planetary Space* was a seminal work that single-handedly, in the words of one historian, "set off an explosion of interest . . . that soon turned into a near frenzy of creativity."

Oberth and his followers quickly learned what Goddard had already realized, that government money for space exploration barely existed. As European leaders recognized the rocket's potential as

Wernher von Braun (1912-1976) built the rockets that sent Americans to the Moon, but first he had to use his charisma and vision to persuade people that the journey could be made.

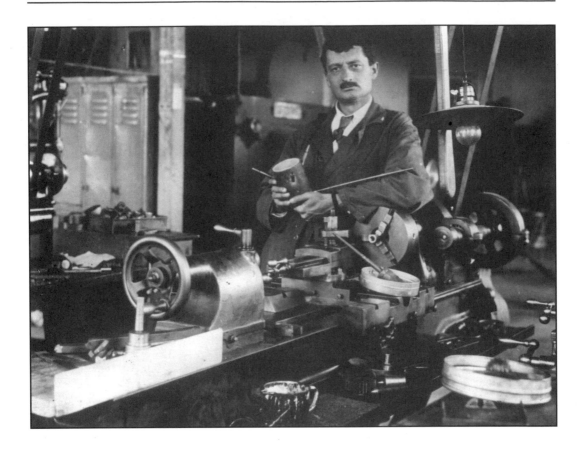

Hermann Julius Oberth (1894-1990) in his laboratory. Oberth mentored von Braun, then later assisted his student in developing rockets in the United States.

a weapon, however, the purse strings loosened. This was especially true in Germany, which had lost World War I (1914-1918). The Treaty of Versailles, which officially ended the conflict, forbade the Germans from owning long-range bombers and cannon—that is, weapons they could use to attack other nations. But the treaty said nothing about rockets, which were undeveloped and therefore not seen as a threat. Germany exploited this loophole and supported rocket development more aggressively than any other country.

No one profited from this more than Wernher von Braun, Oberth's most brilliant disciple. Von Braun most wanted to build rockets to travel to other planets, but only his development of the world's first ballistic missile—the V-2—enabled him to build any rockets at all. After World War II, when von Braun found himself in the United States, depressingly little would change. Again he would build rockets. And again, though several would send men to the Moon, they would mainly be tools of war.

ballistic missile: a rocket that flies under power to a predetermined altitude and location. The engine then cuts off and the missile falls freely to its target.

Wernher Magnus Maximilian Freiherr von Braun was born on March 23, 1912, in Wirsitz, Prussia (part of Poland after 1918). Magnus von Braun, Wernher's father, was a wealthy landowner who moved comfortably within elite government circles. He would serve the post-World War I German government as minister of agriculture. Wernher's mother, Emmy, well educated and conversant in six languages, had a passion for astronomy. She gave him a telescope when he turned 13, igniting the boy's interest in the stars and space travel.

Thus began von Braun's 20-year rise to become the world's foremost rocket pioneer. The first important year was 1925, when 13-year-old Wernher read Hermann Oberth's book, *The Rocket into Planetary Space*. It was then that he committed his life to building a spacegoing rocket. In 1932, von Braun's genius drew the attention of the German army, which recruited the 20-year-old to lead a team to develop a long-range ballistic missile. Adolf Hitler rose to power the following year and eventually launched von Braun's creation, the V-2, against

Only about 20 years old in this photograph, Wernher von Braun (right) carries a Repulsor rocket to be tested. The scientist on the left is Rudolf Nebel, who also worked with Oberth.

the Allies during the last years of World War II. The V-2 bombardment of Britain helped awaken the Allies to the potential of the rocket as a long-range weapon. When von Braun and his team surrendered to U.S. forces in 1945, they were whisked to the States to help rear America's infant rocket program. There were Americans building rockets at this time, namely at the Jet Propulsion Laboratory (JPL) in Pasadena, California. Because Robert Goddard had failed to fully share his work, however, America's homegrown rockets were less advanced than von Braun's.

Although he had never been to America, von Braun found himself on familiar ground. He wanted to build rockets for space flight but found he could not; space travel was still science fiction, too fanciful for government funding. The United States was inching toward war with the Soviet Union, however, and long-range explosive rockets would be a potent addition to the U.S. military. So von Braun built missiles, instead.

From 1945 to 1950, von Braun taught the Americans by firing confiscated V-2s at the White

The German V-2 rocket, shown here on its launching ramp, lived up to its name. The V stands for Vergeltung, *the German word for "retaliation."*

The first Redstone rocket lifts off at Cape Canaveral, Florida, on August 20, 1953.

Sands, New Mexico, test range. The German team then transferred to Redstone Arsenal near Huntsville, Alabama. There von Braun led development of Redstone, America's first medium-range ballistic missile. Redstone (grouped with three upper-stage JPL rockets) went on to launch the first U.S. satellite, *Explorer 1*, in January 1958 and the first astronaut, Alan B. Shepard, in May 1961.

While these were great feats of engineering, they were not Wernher von Braun's alone. "Rocket and spacecraft development," he had written in 1954, "is utterly dependent upon good teamwork." Von Braun was a master at tapping sources—universities for ideas and people, government for funding and facilities, industry for manufacturing. To von Braun, the "team" was more than his elite group of engineers and technicians at Redstone Arsenal (renamed the George C. Marshall Space Flight Center in 1960); it included everyone involved with his work.

The most important team member was the American public. The people paid the bills, and unless they supported his efforts, von Braun knew he would fail. Throughout the 1950s, he became the nation's most recognized and popular advocate for space exploration. He spoke at symposia, hosted a three-part television series on space travel for Walt Disney, and wrote prodigiously—more than 500 essays and several books during his years in the United States. His most influential work went to press in March 1952. The article, "Crossing the Last Frontier," launched a two-year series of articles published by *Collier's* magazine to convince

Americans that space flight was no longer the stuff of science fiction.

Introduced by the editors as "the foremost rocket engineer in the world today," von Braun offered a blueprint for an American space program. The vision centered on a ring-shaped space station that would serve as a platform for scientific study and as a "springboard" for exploration of the Moon and Mars. To carry the outpost to orbit in prefabricated chunks, von Braun proposed a fleet of huge rockets. With a few modifications, they would fly all the way to the Moon.

Von Braun realized money for space hardware would be scarce. He wasn't surprised; he had faced as much in Germany at first. War had come to his rescue then, prompting huge outlays for rocket-powered weapons. Von Braun hoped the Soviet threat would have the same effect now. The Soviets, warned *Collier's*, could build space stations, too, from which they could drop atomic bombs on "any target on the earth's surface with devastating accuracy." Von Braun joined in the argument: Because a space station would pass over all nations, he wrote, it would offer its builders "the most important tactical and strategic advance in military history."

Americans loved the *Collier's* articles. They were fascinated by the idea of space flight, if not yet concerned enough to fund it. As the decade passed the midpoint, von Braun continued to build his rockets and to hope something big would come along to kick-start Americans into supporting space travel. In October 1957, that something big happened.

This man [Wernher von Braun] could convince anybody. His dreams, his ideas are mesmerizing. He is so effective that he could sell anybody anything. Even used cars!
—Cornelius Ryan, von Braun's editor at *Collier's*

THE BREAKTHROUGH

On October 4, a Soviet R-7 intercontinental ballistic missile (ICBM) launched the first artificial satellite, *Sputnik*, into orbit. The event shocked Americans—it suggested that the Soviets held a technological lead over the United States, especially in long-range missiles. Practically overnight, the need to reassert U.S. primacy in rocketry became paramount. In August 1958, von Braun received the go-ahead to develop his super booster. When the Soviets scored another spectacular "first" by orbiting Yuri Gagarin in April 1961, the Americans scrambled to find some race they could win. In May, President John F. Kennedy committed the nation to "landing a man on the Moon and returning him safely to Earth" by the end of the decade. The whole plan revolved around von Braun's heavy-lift rocket, which he called Saturn 5.

At the Marshall Space Flight Center (MSFC) in Huntsville, Alabama, Wernher von Braun immersed himself in every detail of Saturn's creation. As director, he divided MSFC into eight laboratories, each with duties like developing and testing rocket bodies and engines, using wind tunnels to simulate wind flow around airborne rockets, developing instruments for guidance and control, and so on. He didn't try to manage every aspect of development; that would have been impossible. He picked good lieutenants and let them lead. But von Braun did demand final say on all matters. Significantly, he insisted that the industrial contractors who actually built the rockets did so from plans and prototypes

President John Fitzgerald Kennedy (1917-1963) challenged the nation to go to the Moon—and to get there first.

developed by MSFC. Normally, contractors designed their own products from government specifications.

Saturn 5 was a monster. At 363 feet tall, it stood almost 100 feet taller than first imagined by von Braun in *Collier's*. Not surprisingly, the rocket's immense size required many innovations in design and construction. Two in particular stand out. The first involved turbopumps in the five F-1 engines that powered the booster's first stage. Still the most powerful engines ever built, together the F-1s consumed almost 560,000 gallons of propellant (kerosene and liquid oxygen) during the 161 seconds they burned during flight—or about 3,500 gallons every second. Each engine had a turbopump the size of a refrigerator to drive the liquids into the combustion chamber. Not only did the blast of combustion and their own internal friction subject these pumps to temperatures near 1500°F, but the pumps also had to endure the super-chilled -300°F grip of liquid oxygen. It took new metals and manufacturing techniques and hundreds of tests to perfect the pumps.

The other breakthrough involved welding together huge 2,200-pound sheets of aluminum to make the first stage's propellant tanks. The tanks were 33 feet wide and a pair had over six miles of welded seams. To get a perfect seam—and every centimeter was X-rayed to make sure it was perfect—new types of welding machines traveled along specially rigged precision-mounted tracks, doing their job in an area where temperature and humidity were tightly controlled. Joining the pieces sometimes caused ripples in the big sheets, so

Wernher von Braun checks a welding device while touring the manufacturing and engineering laboratory at MSFC in 1967.

von Braun's welders invented an "electromagnetic hammer" that used a magnetic field to "pound" the deformed metal back into shape.

The first Saturn 5 was ready for testing on November 9, 1967. As the unmanned rocket rose above its launch pad at Cape Canaveral, Florida, atop a tail of yellow-white flame as long as itself, its first-stage engines produced 180 million horsepower,

about twice the electric power that could be generated by harnessing every river in North America. Pedestrians viewed the ascent from the streets of Jacksonville, Florida, 150 miles away. Much closer, about three and half miles from the launch pad, reporters stared in awe. They couldn't hear the rocket—the sound wave hadn't reached them yet— but they could feel it shake the ground. Then, 16 seconds after it started, the overwhelming sound of ignition struck. One journalist likened it to a continual "thunder . . . overlaid with the roar of a dozen hurricanes." The first Saturn 5 went on to perform nearly flawlessly.

Saturn 5 number two lifted off five months later, in April 1968. It was anything but flawless. Misfirings in the first-stage F-1 engines sent violent shudders through the unmanned command module attached above the booster's third stage. The second stage performed even worse, as two of its five J-2 engines quit altogether. Finally, the single J-2 that powered the third stage also failed. It was clear that von Braun's people had more work to do.

They traced the shudder in the F-1s to the surging action of propellant through the liquid oxygen lines and quickly fixed the problem. The mystery of the malfunctioning J-2 engines was just as speedily solved—their propellant lines were adequate for ground firing, but thin enough to fail in airless space. With the addition of thicker fuel lines in the J-2s, von Braun convinced NASA to rate the Saturn 5 rocket ready to transport humans.

command module (CM): the section of an Apollo spacecraft containing the crew's living area and the instruments and equipment they need

The first piloted Apollo mission was actually Apollo 7 *in October 1968, but the three astronauts on board rode into space on a two-stage Saturn 1B rocket, not the more powerful three-stage Saturn 5 used by all subsequent missions to the Moon—and pictured here.*

Number three not only carried astronauts, it also launched them all the way to the Moon. Over Christmas 1968, Frank Borman, James Lovell, and William Anders flew the historic mission, *Apollo 8*. They didn't land—they had no lunar lander, which was not yet ready—but their booster worked perfectly to deliver them to the point in lunar orbit where future crews would begin their descents. One of those, the crew of *Apollo 11*, would be the first to go all the way to the Moon's surface.

THE RESULT

On July 20, 1969, *Apollo 11* astronauts Neil Armstrong and Edwin "Buzz" Aldrin climbed inside their lunar lander, undocked from the command module being held by Michael Collins in orbit around the Moon, and descended to the surface. Several hours later, Armstrong and Aldrin became the first men to walk the lunar surface.

Neil Armstrong snapped this photo of Buzz Aldrin after Aldrin joined him on the surface of the Moon. Their landing module and Armstrong are both reflected in Aldrin's visor.

The event remains the crowning achievement of the Space Age. Without von Braun, the walk could not have happened; the new age would not have begun. In fact, the blueprint he unveiled in *Collier's* still guides NASA. The super booster he envisioned in 1952 was reusable, with a winged third stage that could fly back to Earth like an airplane. That sounds a lot like today's space shuttle. And the space station he mentioned? Well, *Alpha*, the international space station, isn't donut-shaped, but it's up there. As for the rest of von Braun's dream, to use the space station as a launch platform for manned missions to Mars, only time will tell. It took a world war to reach the edge of space and a cold war to explore the Moon. What will push us to walk on Mars?

Wernher von Braun, who owed his success to international conflict, now watched as a lull in the Cold War killed his super booster. Americans believed they had won a major victory by beating the Soviets to the Moon. The space race was clearly over. On the terrestrial front, the Vietnam War was winding down. (U.S. involvement in the war would end in 1973.) The early 1970s was a time of détente, a "lessening of tensions" with both the Soviet Union and Communist China. In short, Americans lost the reasons to fight, and with them the will to fund big-budget space travel. The last three Moon missions, *Apollo 18*, *19*, and *20*, were scrubbed, and their Saturn 5 rockets relegated to use as tourist attractions. One, lying in a grassy field at the Johnson Space Center in Houston, Texas, is home to owls.

The huge first-stage engines of a Saturn 5 rocket dwarf their developer, Wernher von Braun.

With the Moon program winding down and hopes for a follow-up to Mars dashed amid budget cuts, von Braun left NASA in June 1972 to become vice president of engineering and development at Fairchild Industries, an aerospace company in Germantown, Maryland. In December 1976, he was diagnosed with cancer. He died the following year on June 16. His stature as space flight visionary and pioneer has not been reached by anyone since.

CHAPTER THREE

Max Faget
and the Space Capsule

Wernher von Braun had a knack for launching astronauts into space. Of course, he didn't just strap men to rockets and fire the engines. Astronauts flew in spacecraft. But while the basic shape and action of rockets has been set since their inception a millennium ago, spacecraft are a modern invention. As need for them arose in the late 1950s, it wasn't at all clear what form they would take or how they would work.

This changed as engineers took a hard look at the rockets themselves. All were derived from ballistic missiles. A ballistic missile works by flying to a predetermined point above the atmosphere and releasing its explosive nose cone, which continues up and over in a great ballistic arch. To hit a target, the explosive tip must do nothing but fall like a stone and survive reentry. It is terribly effective despite its simplicity. Then again, the nose cone works so well because it is so simple. When Max Faget pondered

Maxime Allan Faget (b. 1921). "If it is possible to trace the inspiration and inventiveness behind the U.S. space program to a single individual, Faget would be a likely source," wrote Kenneth A. Brown in Inventors at Work. *Faget holds numerous patents for elements of the Mercury capsule as well as the space shuttle.*

designs for America's first manned spacecraft in 1957, he considered the uncomplicated ballistic nose cone. If the cone could deliver a nuclear bomb through space, could it not be modified to transport a human?

Guy Faget was a doctor with the U.S. Public Health Service and stationed in British Honduras when his second son, Maxime Allan, was born there on August 26, 1921. Guy Faget held subsequent posts in the United States. As head of the national "leprosarium" in New Orleans, Louisiana, in the mid-1940s, he discovered the first effective cure for leprosy, an infectious disease of the skin. Nor was Max's father the first innovative Faget. Guy's grandfather, Jean-Charles Faget, also a physician, was first to identify and list the early symptoms of the tropical infection yellow fever.

Max's own extraordinary journey began as a child. He and older brother Frank read airplane magazines and built models. "I was a slapdash, trial-and-error type," he recalled, interested more in achieving solid designs than fancy ones. Both brothers decided early in their lives to become aeronautical engineers.

After graduating from George Washington High School in San Francisco, California, in 1939, Faget studied for a year at an area junior college before transferring to Louisiana State University in Baton Rouge. In December 1941, America entered World War II. Hoping to fly for the U.S. Navy, Max joined the university's reserve officer training corps. He graduated in 1943, a naval officer with a degree in mechanical engineering.

On active duty, Faget ended up about as far as he could get from flying—in a submarine. But that was okay; he and Frank had built model subs, too. "A submarine is a very high-tech ship," he explained, "very compact, and full of machinery, like a spacecraft." Faget's sub spent much of the next two years running combat patrols against Japanese surface ships.

After the war, in summer 1946, Faget journeyed east to the Langley Aeronautical Laboratory in Hampton, Virginia, to find work. Part of the National Advisory Committee for Aeronautics (the NACA, precursor organization to NASA), Langley was one of the premier research facilities in the country. Robert Gilruth, chief of Langley's Pilotless Aircraft Research Division, was impressed with Faget's enthusiasm and hired him on the spot. Faget went to work designing high-speed jet engines called ramjets. In 1950, two of his ramjets powered an aircraft to 2,200 miles per hour and 65,000 feet altitude, both records at the time for airplanes with air-breathing engines. Faget remained at the forefront of aviation as he and three others drafted the preliminary designs for the X-15 rocket-engine plane in 1954. The NACA and the U.S. Air Force used the vehicle to experiment with low-level space flight.

In 1957, when Faget and others considered designs for the next X-plane, they thought about a bomber that could fly 10,000 mph. That was well below orbital speed of about 17,500 mph, but few people from President Dwight Eisenhower down really cared about getting to orbit. This changed after October 4, when the Soviet Union launched

The seal of the National Advisory Committee for Aeronautics (pronounced as letters N-A-C-A, not as a word), established by Congress in 1915

Robert R. Gilruth (1913-2000)

Sputnik, the first artificial satellite. Eisenhower never considered *Sputnik* a threat, but he bowed to public pressure and ordered the U.S. Army to match the Russian feat as soon as possible. Using a multistage ballistic missile built by teams under Wernher von Braun and William Pickering, the army launched *Explorer 1* on January 31, 1958.

In the months between *Sputnik* and *Explorer 1*, Faget put together a group to investigate the next giant step in space flight—launching a human into orbit. It became obvious that if America was to beat the Soviet Union, it could not spend time developing ever-faster X-planes. The U.S. would have to use a ballistic missile. The tough question then became how to return the orbiting pilot safely to Earth.

There were two schools of thought within the NACA on how best to return from space. One group, led by Alfred J. Eggers Jr. of the NACA's Ames Aeronautical Laboratory in California, thought returning spacecraft should fly back. He did not advocate the use of wings, since anything protruding from the vehicle would be very difficult to keep from burning off during reentry. Eggers did think it was possible to shape the spacecraft's body so it would generate lift and fly without wings. One advantage of "lifting" reentry was entering the atmosphere at a relatively shallow angle. This would help the crew, who would face slower deceleration from orbital speed. Another advantage to lifting reentry was the possibility of landing horizontally on a runway. This meant that the crew could land in more than one location and that the spacecraft could be reused.

A simpler method was championed by Eggers's boss at Ames, Harvey Allen. Several years back, Allen had conceived of the "blunt-body" reentry vehicle. At the time, conventional thinkers considered pointed shapes best for reentry, just as they were for high-speed aircraft. But speed, thought the unconventional Allen, was precisely the problem. The atmosphere is composed of air molecules that are packed tighter together the closer they are to the ground. The faster something moves through the dense lower atmosphere, the hotter it gets. Allen thought it would be better to slow an incoming object quickly, before it reached the lower atmosphere. For this he proposed a blunt shape, which has lots of drag.

Faget agreed. He feared Eggers's design would accumulate too much heat during its extended descent through the atmosphere. A plunging blunt body, on the other hand, would push a "shock wave" in front of it as it fell that would absorb and dissipate much of the heat. An even more practical concern was size. Faget knew blunt bodies would be small enough and light enough to ride current ballistic missiles because they already did so as explosive warheads. Eggers's lifting body would be too heavy. Finally, there was the issue of complexity. Because a blunt body craft would simply drop into the ocean (under a parachute, of course), an astronaut could survive even if he became incapacitated during reentry. Faget made his choice. By the end of March 1958, he had convinced the NACA to adopt the ballistic nose cone as the basis for manned space flight.

H. Julian Allen (1910-1977), known to most of his colleagues as "Harvey," describing his blunt-body theory in 1957

blunt body: a spacecraft with a flattened, or rounded, nose designed to slow the craft down and reduce heat during reentry into Earth's atmosphere

lifting body: a wingless aircraft that generates lift with the shape of its body alone

THE BREAKTHROUGH

The Faget team's first innovation was to design the cone-shaped space capsule to flip around just before reentry and descend broad bottom first. This would increase drag and the effectiveness of the heat-dissipating shock wave. Further protecting the astronaut from the 2,500° F plunge would be a fiberglass-composite heat shield affixed to the capsule's bottom. The material of the shield was designed to ablate (or burn away) during reentry, taking the extreme frictional heat with it. Deorbit would be initiated by a retrorocket strapped to the outside of the heat shield. When fired in the direction of travel, it would slow the craft and drop it from orbit.

Many innovations involved crew safety. In the event of a disaster on the launch pad—say, if the booster exploded beneath the spacecraft—the astronaut would need a quick getaway. One of Faget's people earlier had designed rockets for towing aircraft models with a cable for high-speed tests. Faget thought tractor rockets could pull the space capsule up and away from an exploding booster. He designed a rigid tower to sit atop the capsule with its tractor rockets on top. If the tower wasn't needed, it would be jettisoned minutes after liftoff.

High g's, during reentry but especially during an "off-the-pad" abort, posed a real danger. This concerned the design team. Tests by the Germans during World War II had shown that the best position for enduring high g's was for a pilot to lie on his back. But would position alone offer enough protection? Then Faget thought about eggs: "I had read somewhere that if you properly cradle an egg, you can subject it to very high g's, and the inside of the egg doesn't get hurt." Faget's idea was to support the supine pilot with a formfitting couch, as if he were an egg in a carton. In July 1958, test subject Carter C. Collins, nestled snugly in a Faget couch, rode the NASA centrifuge at Johnsville, Pennsylvania, to 20 g's, slightly more force than an astronaut might endure during an abort and more than twice that expected during normal reentry.

centrifuge: an apparatus in which humans or animals are enclosed and which is revolved to simulate space flight acceleration

In 1959, technicians fit a couch destined to be installed in a Mercury space capsule.

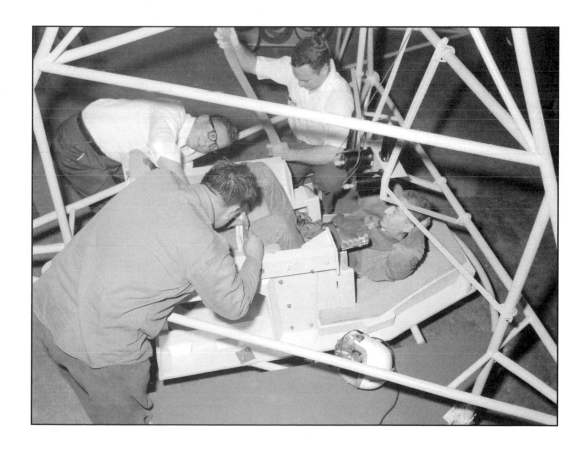

Thomas Keith Glennan was NASA's first administrator, from 1958 to 1961. Under his leadership, NASA absorbed the NACA, Langley and Ames Aeronautical Laboratories, and several other programs to become a strong federal agency.

During July, President Eisenhower signed a law transforming the NACA into NASA, the National Aeronautics and Space Administration. He ordered the new agency to send a man into orbit. To do the job, NASA chose Faget's design over others proposed by the military. NASA administrator Thomas Keith Glennan named the project "Mercury." Glennan formed the Space Task Group to implement Mercury and placed both under Faget's boss, Robert Gilruth.

Meanwhile, development of the spacecraft continued rapidly. To test the capsule's escape tower and other systems, Faget and a teammate designed a squat, five-story-tall rocket booster they dubbed Little Joe. "Boilerplate" capsules (stripped-down mockups) were tested on Little Joe and any other boosters Faget could scrounge, including several ballistic missiles purchased from the army and air force. The team dropped mockups from airplanes and balloons to fine-tune the landing parachute and tested model capsules in wind tunnels to ensure stability at anticipated altitudes and speeds.

By spring 1959, Project Mercury was humming along. In April, Eisenhower granted the project the nation's highest procurement rating. Now, whatever the Space Task Group needed, it received. Some 4,000 industrial suppliers from 25 states soon contributed parts. Harvey Allen's vision of a blunt-body spacecraft was becoming real, thanks to government funding and the brilliant ideas of Faget and his engineers.

The performance of the Mercury space capsule (above) is studied in the full-scale wind tunnel at Langley Research Center in 1959, one of many tests performed on the craft. At left, a Little Joe booster rocket launches a test capsule.

THE RESULT

Of course, no "manned" spacecraft would be complete without the men. On April 9, 1959, the "Mercury Seven" astronauts were introduced to the press. They were among the elite of their nation's military flyers. But, then, so were the six cosmonauts training inside the Soviet Union for their nation's first space flight. On April 12, 1961, it was one of the Soviets, cosmonaut Yuri Gagarin, who rode inside a spherical Vostok capsule and became the first person in space. Gagarin beat by three weeks astronaut Alan Shepard, who made America's first space flight on May 5 aboard Mercury capsule *Freedom* 7.

Although it placed second to the Russians, the Mercury spacecraft worked splendidly. Shepard's success emboldened the new president, John F. Kennedy, to challenge the Soviet Union to a more difficult race—to land a man on the Moon and bring him back by the end of the decade. The new project was called Apollo.

Actually, NASA had been considering lunar missions since 1959. By October 1960, Faget and company had basic designs in hand for a three-man moonship. The conical shape they chose was Mercury-esque. The spacecraft eventually included three parts: the command module, or the capsule the astronauts would ride in to and from lunar orbit; the service module, the cylinder attached to the capsule's base that would house water and oxygen for the crew and electricity and propulsion systems

Russian cosmonaut Yuri Alexeyevich Gagarin (1934-1968) orbited Earth once, traveling through space for 108 minutes on his historic journey.

American astronaut Alan B. Shepard Jr. (1923-1998) stands in front of Freedom 7 *shortly after his brief suborbital (not fast enough to achieve orbit) flight into space.*

for the spacecraft; and the lunar module, the bug-shaped lander that would ferry two of the crew to the Moon's surface and back.

Faget oversaw development of all three, and this time the money flowed even more freely. "Through its NASA contracts," notes one historian, "the United States Government mobilized a large segment of American industry comparable to that ordinarily achieved only during the exigencies of war." At its height, the Apollo project employed some 8,000 firms and 150,000 engineers and scientists from government and industry. "NASA recruited people as if it were building an army for a war," Faget affirmed.

The successes began with the *Apollo 7* mission in October 1968, when astronauts tested the command module while in orbit around Earth. The high point of the project—and, to date, of the entire Space Age—came on July 20, 1969, when astronauts from *Apollo 11* left the lander to explore the lunar surface. Five more crews followed in their footsteps through December 1972.

NASA officials, including Max Faget (left), and flight controllers celebrate the successful recovery of the Apollo 11 astronauts.

Government support for the space program, however, began to wither even before the first Moon landing. NASA's budget for 1972 was only two-thirds the total allocated for 1966, Apollo's peak year of funding. In the middle of this downturn, Bob Gilruth directed Faget to design the next-generation spacecraft. What would become the space shuttle started out as a wood-and-paper model not much different from the ones Max and Frank Faget had built as kids. From there Faget's team wrote the book on the shuttle—what it would weigh, how powerful it would be, how it would land, and so on. But NASA's chief designer didn't receive everything he wanted. So acute was the agency's budget situation that it had to find additional funding from the U.S. Air Force. Although the military demanded several design alterations, the shuttle that first flew on April 12, 1981, was very much the Faget team's spacecraft.

Faget retired from NASA in 1981. The following year, he took control of a private company developing a low-cost commercial space station. The outpost was to be simple in design, a Faget trademark, and small enough to ride to space in the shuttle's payload bay. The company hoped to lease out space in the station to others, including NASA itself. In the end, the concept never flew and the space agency pushed ahead with plans to build the international station *Alpha*.

Today, Max Faget is retired and lives in Clear Lake City, Texas, just south of Houston. His home is but a stone's throw from the Johnson Space Center, NASA's human space flight center.

payload bay: the area of the space shuttle used to carry cargo, located behind the crew compartment and in front of the engines. In orbit, double doors in the payload bay can be opened to deploy, or release, cargo.

CHAPTER FOUR

John Houbolt
and Lunar Orbit Rendezvous

In summer 1960, as work on the Saturn rocket and Apollo spacecraft progressed, two views predominated within NASA on how astronauts might one day use the machines to reach the Moon. Some favored Wernher von Braun's idea—Earth orbit rendezvous, or EOR. This method involved combining the several modules of the moonship in Earth orbit from parts brought up by two or more Saturn rockets, then flying the whole to a lunar touchdown. Max Faget and the Space Task Group preferred a straight shot to the Moon. They wanted to launch the entire ship from Earth atop a proposed super Saturn called Nova. NASA headquarters favored this approach, called direct ascent or direct flight.

For several months, a group of NASA engineers had been developing a third option—lunar orbit rendezvous. LOR, they insisted, would be safer for astronauts to perform, as well as cheaper and quicker to implement. Indeed, they would

Aircraft designer John C. Houbolt (b. 1919) devised a method to reach the Moon, but first he had to convince his colleagues that the plan would work.

Earth orbit rendezvous (EOR): a proposed method of landing on the Moon by combining several parts of a moonship in Earth orbit, then flying the whole ship to the Moon. Pronounced as letters E-O-R, not as a word.

lunar orbit rendezvous (LOR): the method NASA chose to send astronauts to the Moon, requiring a three-part spacecraft to be launched from Earth to lunar orbit. Pronounced as letters L-O-R, not as a word.

glider: a heavier-than-air aircraft that flies with no engine

argue, if the United States hoped to reach the Moon by 1970, NASA had no choice but to embrace LOR.

Von Braun, Faget, and most of NASA dismissed lunar orbit rendezvous out of hand. A battle of ideas ensued. Heading the noisy group pushing the lunar orbit method was John Houbolt. Houbolt had spent years studying aircraft design. He knew the devil was in the details of the others' plans—in this case, the massive moonship that both EOR and direct ascent demanded. Houbolt was sure it could not work. Luckily for the group and ultimately the U.S. space program, Houbolt was a passionate man who refused to quit until he got what he wanted.

John Cornelius Houbolt was born in Altoona, Iowa, on April 10, 1919. Parents Henrika and John were farmers. Like most farm kids growing up during those years, young John worked long hours in the fields. He developed a capacity for hard work that served him well in later life.

John discovered model airplanes early on. He scrounged what wood and glue he could, and before long his feather-light gliders were riding currents of warm air called "thermals." Poring over second-hand copies of magazines like *Popular Mechanics* helped the boy understand how and why forces like wind and heat affected his models. With that knowledge, John built bigger and better airplanes. One, with a small engine and a five-foot wingspan, rode a thermal for 20 miles!

In spring 1936, John graduated from Joliet Township High School. (The family had moved to a farm in Joliet, Illinois, in 1925.) Two years at

Joliet Junior College followed. Houbolt excelled and won a scholarship for his final two years at the University of Illinois, where he earned a bachelor's degree in civil engineering in 1940 and a master's degree in 1942.

During summers at Illinois, Houbolt worked as an engineer, designing roads and sidewalks for the nearby city of Waukegan and bridges for the Illinois Central Railroad. The latter particularly interested him. As strange as it might seem, the anatomy of a reinforced bridge is similar to that of an aircraft wing. The first uses concrete or steel, and the second, aluminum, but both include spans that must be properly joined and supported. When Houbolt accepted a position in aircraft design at the NACA's Langley Aeronautical Laboratory in Hampton, Virginia, in June 1942, he felt it was a continuation of what he'd been studying practically his whole life.

In 1949, Houbolt became assistant chief of Langley's dynamic loads division, where he directed research into the effects on aircraft of stresses like turbulence, heating (caused by high speed), and landing. The same year he spent six months as an exchange research scientist at the Royal Aircraft Establishment in Farnborough, England. In 1956, Houbolt won the Rockefeller Public Service Award, an honor bestowed yearly by the federal government upon its most outstanding civilian employees. The terms of the award allowed Houbolt to study anywhere in the world with full pay for a year. He chose the renowned Swiss Federal Institute of Technology in Zurich, Switzerland. John Houbolt returned to the United States in 1958 with a doctorate (Ph.D) in technical sciences—a degree that usually takes two to three years to earn.

By this time, the United States was in a furious race with the Soviet Union to send a man into space. (The Soviets had launched the race with *Sputnik* in 1957.) Houbolt was as captivated as anyone with the idea of space flight. But from the start he looked beyond the initial challenge of orbiting a single-occupant spacecraft to more complex missions, like launching a space station. Since a station would be too massive to orbit in one shot, NASA would have to send it up in pieces. The several parts would then rendezvous and dock, or join. The challenge of orbital rendezvous intrigued Houbolt. As he and a few colleagues at Langley delved deeper into the subject, they realized that rendezvous was the key to an even greater endeavor NASA began considering around spring 1959—exploring the Moon.

The idea Houbolt and friends developed was to launch a spaceship to lunar orbit. Once on station around the Moon, a small part of the whole ship would descend to the surface. The lander's crew would explore, then return to orbit to rendezvous with the mothership. With the astronauts again united, the mothership would return to Earth. Houbolt and company called the technique "lunar orbit rendezvous."

"The reception of our LOR proposal [within NASA]," Houbolt recalled, "was discouraging and distressing." Max Faget wanted nothing to do with tricky rendezvous maneuvers; astronauts had yet to

rendezvous: the close approach of two spacecraft in orbit. The two craft stay close together and sometimes dock with each other.

dock: to join two or more spacecraft in space

Rendezvous is a vital maneuver not only for orbital missions involving people, but also in all space endeavors. Even a simple communications satellite, for example, must rendezvous with a specific location in space.

fly in space, and he wasn't confident that they could pilot a rendezvous when they finally did. He preferred the launch-pad-to-lunar-surface method called direct ascent.

Wernher von Braun preferred Earth orbit rendezvous, which involved building a moonship in Earth orbit—he dreamed of a space station and thought the exercise would be great practice. But he shared Faget's anxiety regarding complex maneuvers at the Moon, which was a quarter-million miles from home and too far to effect a rescue should a disaster

John Houbolt presenting his lunar orbit rendezvous method of landing on the moon. His diagram clearly shows the lunar lander's separation from the mothership while in orbit around the Moon, its descent to the Moon (6), then ascent (7) and docking with the main craft (8). The lander would be left behind in space once the astronauts were safely transferred.

strand the astronauts. Like Faget, von Braun thought the best solution would be to fly the entire moonship to the lunar surface. It was this aspect of von Braun's and Faget's schemes—a massive moonship projected to weigh up to 130,000 pounds—that Houbolt's group most objected to.

The first problem with a ship this big would be just getting it to the Moon. To break free of Earth's gravity, it would have to fly about 24,500 miles per hour. The more massive the payload boosted to this speed (known as earth-escape velocity), the more power required to boost it. This is why EOR needed at least two Saturn rockets, and direct ascent needed the super Nova. Landing the ship on the Moon would be an even bigger challenge. The astronauts would have to back the ship—the height of a six-story building—to the ground. If they lived through that, they would then face a 60- to 70-foot climb down to the lunar surface in bulky space suits.

These obstacles disappeared with LOR. Only a simple lander, projected to weigh around 19,000 pounds and to stand 14 feet, would descend to the surface. A few years later Houbolt wrote:

> I can still remember the 'back of the envelope' type of calculations I made to check that the scheme resulted in a very substantial savings in Earth boost requirements. Almost spontaneously, it became clear that lunar orbit rendezvous offered a chain reaction simplification on all back effects: development, testing, manufacturing, erection, count-down, flight operations, etc. . . . All would be simplified.

velocity: the rate at which a body moves in a certain direction, expressed in measures of distance and time, for example, miles per hour in a straight line. Speed is the rate of motion regardless of direction.

space suit: a protective garment with life-support and communication systems that enable the wearer to survive in space

From that moment in summer 1960, Houbolt became the crusader for LOR. He was convinced the plan was the simplest, fastest way to reach the Moon. Unfortunately for him, von Braun, Faget, and NASA headquarters didn't see things the same way. As one historian has noted, they made the next few years "the most hectic and challenging period of John Houbolt's life."

This massive five-stage Nova launch vehicle would have been necessary to send a spacecraft directly to the Moon's surface. With LOR, the three-stage Saturn 5 rocket proved sufficient.

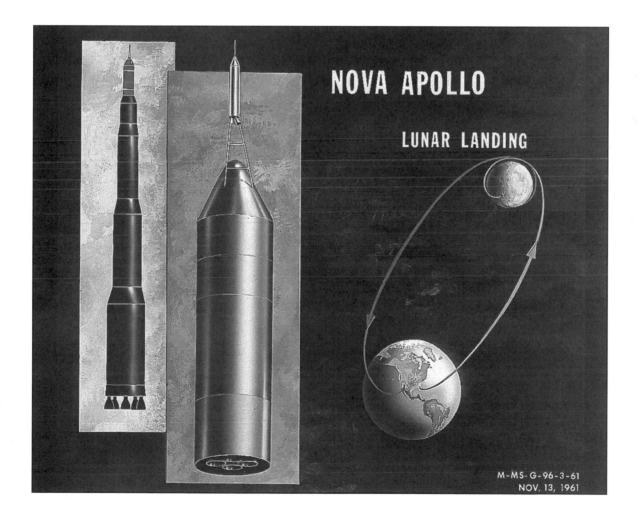

NOVA APOLLO

LUNAR LANDING

M-MS-G-96-3-61
NOV. 13, 1961

THE BREAKTHROUGH

In late May 1961, President John Kennedy committed the United States to landing a man on the Moon and returning him safely to Earth by the end of the decade. It was a bold stroke, and not only because astronaut Alan Shepard had made America's first space flight just three weeks earlier. Kennedy's move was audacious because NASA wasn't sure how to do the job.

The space agency's indecision was no fault of John Houbolt's. At Langley in September 1960, he championed the benefits of lunar orbit rendezvous to Robert Seamans, who ran NASA's day-to-day operations as associate director. Houbolt's passion impressed Seamans. He invited the engineer to present his ideas to von Braun, Faget, and others at headquarters. As Houbolt finished his lecture that day, some in the crowd considered his analysis of the weight saved by using the LOR system too optimistic. The outspoken Faget even denounced Houbolt's figures as "lies."

Houbolt pressed on. He knew his numbers on weight were right. Furthermore, his calculations showed that lunar orbit rendezvous promised the highest probability of success in every phase of a mission. He argued his case again before a two-day NASA-wide meeting in January 1961. A week later, he and three colleagues met with representatives of the Space Task Group. Another briefing followed at headquarters the same month. Still another presentation before the group came in April.

Robert Seamans (left) tours Cape Canaveral with Wernher von Braun and President John Kennedy.

While none of these meetings seemed to John Houbolt to bear fruit, he was in fact gaining converts. One new disciple was the Space Task Group member James Chamberlin. After the April meeting, he asked Houbolt for everything he had on rendezvous. In February, Chamberlin had been given the task of developing a follow-up project to Mercury. As early as December 1959, Houbolt had suggested a rendezvous mission with a satellite in Earth orbit to "define and solve" the problems of joining spacecraft. Now, in the April meeting attended by Chamberlin, Houbolt proposed a project called MORAD (Manned Orbital Rendezvous and Docking), which would use an unmanned rocket stage as a docking target for an advanced Mercury capsule. Chamberlin loved the idea and convinced the Space Task Group to approve just such a program. Project Gemini was announced in December 1961. In a series of eight space flights in 1965-1966, the astronauts of Gemini developed many skills they would later use to reach the Moon, including rendezvous and docking with unmanned rocket stages.

But that was the future. In spring 1961, despite the fact that Houbolt had given 16 presentations on the subject of rendezvous since September 1959, NASA seemed no closer to choosing LOR. In May, Houbolt wrote directly to Associate Director Seamans. It was a risky move from a career standpoint, as NASA (like most large organizations) expected its people to work through a chain of command. Houbolt hoped to shake headquarters out of

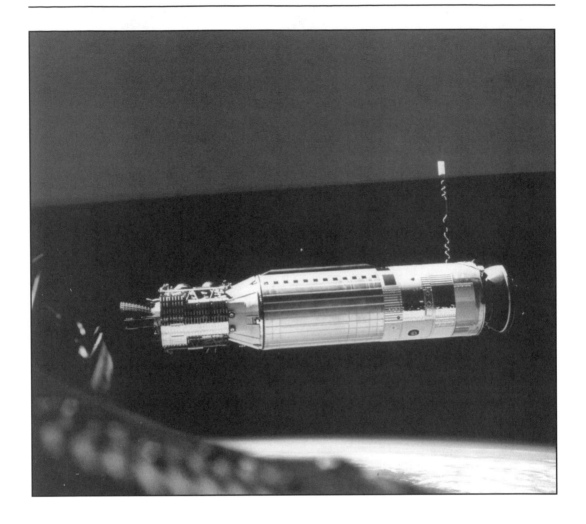

Astronauts Neil Armstrong and David Scott, aboard Gemini 8, prepare to dock with this Agena rocket during their March 1966 mission. John Houbolt proposed the idea that astronauts practice rendezvous and docking in Earth orbit.

its near hypnotic attraction to direct ascent. The Nova booster wouldn't be ready in time, he explained, so it was "mandatory" that NASA include rendezvous in its planning. The bold tactic worked; Seamans created a committee to consider the maneuver.

Houbolt next attacked Earth orbit rendezvous. EOR "was becoming a beast," he later recalled, involving "putting together five pieces of hardware

[in orbit]. It was getting to be a great, big, long
cigar." In the fall of 1961, he and several colleagues
put together a two-volume report explaining why
lunar orbit rendezvous was the quickest, safest
approach. Houbolt included the work in a second
letter to Seamans in November. He pulled no
punches. "Do we want to go to the Moon or not?"
he asked point-blank. Contacting the associate
administrator was "somewhat unorthodox," he
admitted, "but the issues at stake are crucial enough
to us all that an unusual course is warranted." Again
Houbolt pleaded that his "much less grandiose
scheme" be considered. Once again Seamans was
touched by Houbolt's passion and sound reasoning.
He passed the letter along to the Office of Manned
Space Flight with a firm assurance that LOR would
get its due consideration.

This time it did. With Seamans's backing,
LOR became the favorite at headquarters. Faget
and Space Task Group climbed aboard after finally
ditching the idea of a monster moonship. The cli-
max came at a conference in June 1962 at Marshall
Space Flight Center. Von Braun, Marshall's director, over a "storm" of objections from his surprised
staff, abruptly abandoned EOR for Houbolt's
method. The rocketeer still believed that Earth
orbit rendezvous would help with a space station,
but he realized an EOR mission was just too complex
to accomplish by the president's deadline.

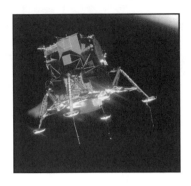

The Eagle (from Apollo 11) in lunar orbit. The lunar module or lander (LM) was designed to carry astronauts from the Apollo command module to the Moon's surface and back.

An Apollo command/service module, orbiting around the Moon. The command module (CM) was the cone-shaped section of an Apollo spacecraft where the crew traveled. The service module contained the spacecraft's main engine, fuel cells, and other supplies.

THE RESULT

The first successful application of lunar orbit rendezvous was by the crew of *Apollo 10* in May 1969. The landing module didn't descend all the way to the surface, but it undocked and went through all the motions short of a touchdown before rejoining the command capsule. Actual landings began with *Apollo 11* in July (five months before the presidential deadline) and continued through *Apollo 17* in December 1972. Of the six Apollo missions that reached the Moon, six landers undocked and landed, six crews walked the surface, and six landers returned to rendezvous with their motherships. LOR worked perfectly, just the way John Houbolt said it would.

There was, of course, one Apollo mission that never reached the Moon. But the near-fatal flight of *Apollo 13* had nothing to do with Houbolt's method; indeed, LOR saved the crew's life. When an oxygen tank exploded aboard the spacecraft on its way to the Moon in April 1970, James A. Lovell Jr., John L. Swigert Jr., and Fred W. Haise Jr. took refuge in the ship's "lifeboat"—the undamaged lunar lander that Lovell and Haise would have piloted to the surface. That the lander could be occupied in an emergency was one of the benefits of LOR that Houbolt had cited years earlier. The forward-looking engineer had seen the future in more ways than one.

Houbolt became chief of Langley's Theoretical Mechanics Division in February 1962, where he oversaw experimental work on guidance and control for aircraft and spacecraft. He left NASA for the private sector the following year. Houbolt returned to Langley in 1976 as Chief Aeronautical Scientist, a post that allowed him to do unfettered research into new aircraft designs. He retired from NASA for good in 1985.

John Houbolt currently lives in Scarborough, Maine, where he has been studying the effects of atmospheric turbulence on airplanes. He works as a consultant to the Federal Aviation Administration on this subject.

Commander Philip Jerauld (at microphone), ship's chaplain for the U.S.S. Iwo Jima, offers a prayer of thanks for the safe return of the Apollo 13 crew soon after they arrived aboard the recovery ship. The astronauts are (from left of Jerauld), James A. Lovell Jr., Fred W. Haise Jr., and John L. Swigert Jr.

CHAPTER FIVE

William Pickering and the Planetary Probe

Without the power of Wernher von Braun's Redstone rocket, America's first satellite, *Explorer 1*, would not have reached orbit on January 31, 1958. But the booster that launched America into the Space Age wasn't your everyday Redstone, and it wasn't all von Braun's. This Redstone had four stages, not one, and was called by a different name: Jupiter-C. While von Braun's team built the booster's first stage, the top three came from the Jet Propulsion Laboratory (JPL) in California. Oh, JPL built the satellite, too.

JPL engineers had long been in the vanguard of American rocket development. In fact, the first artificial object to leave Earth's atmosphere—if only briefly before falling back—was WAC Corporal, a JPL test rocket that flew straight up for more than 40 miles in October 1945. It was the three Explorers launched in 1958, however, that propelled JPL to fame.

As director of the Jet Propulsion Laboratory from 1954 to 1976, William Hayward Pickering (b. 1910) led many exciting projects, from the creation of the first U.S. satellite to the successful Viking missions in the 1970s.

79

Powered by a Jupiter-C launch vehicle, Explorer 1 *lifts off on from Cape Canaveral on January 31, 1958.*

For director William Pickering, the big question became "what next?" Human space flight was intriguing, but Pickering could see how colossal an undertaking that would be. He wanted something smaller, something JPL could manage itself. Earth satellites like Explorer weren't the answer, either. They were small enough, but, frankly, JPL had been there and done that.

The last option was exploring the planets with robot probes. True, space probes were something so new they literally had to be launched from the ground up; but, with the heady successes of Explorer behind him, Pickering was confident his people could overcome the obstacles. Perhaps he was too confident.

William Hayward Pickering was born on December 24, 1910, in Wellington, New Zealand. Son of a pharmacist, William was sent at age four to live with his grandmother in the small fishing village of Havelock after his mother died. There he attended elementary school. Later his father sent him to Wellington College (what we would call a high school) to complete his public education.

Pickering traces his earliest interest in science to radio. In high school, he and a friend built one of New Zealand's first amateur radio stations, assembling most of the equipment from bits and pieces. They used it to communicate with stations as far away as the United States. Astronomy also fascinated William.

For a year after high school, William studied electrical engineering at Canterbury College, now the University of Canterbury. In the spring of 1929, Pickering accepted his uncle's invitation to move to the United States and study at the California Institute of Technology in Pasadena, California.

Pickering excelled at Caltech and caught the eye of Nobel Prize-winning physicist Robert A. Millikan. At Millikan's instigation, Pickering studied extraterrestrial radiation, or "cosmic rays"—a term Millikan coined. Pickering was among the first researchers anywhere to use Geiger counters to measure cosmic rays. He did so by designing instruments to transmit the counters' findings and attaching both to high-flying balloons. This method of transmitting scientific measurements by radio, pioneered here by Pickering, was called telemetry. Pickering went on to earn a Ph.D. in physics and join the Caltech faculty in 1936.

In 1944, Pickering transferred to Caltech's Jet Propulsion Laboratory. Theodore von Kármán had founded the lab in the mid-1930s to build rockets. During World War II (1939-1945), JPL supplied rockets to the U.S. Army Air Corps to help overloaded aircraft take off. The lab expanded into weaponry when army leaders, having learned of the German V-2s, wanted their own missiles. The lab gave the project to Pickering, whose knowledge of electronics facilitated development of the missile's guidance and control.

The result was a missile called Corporal, launched in May 1947. While working on Corporal, Pickering and JPL collaborated with Wernher von Braun's rocket team to produce Bumper-WAC—a

radiation: the giving off of energy in the form of rays or waves. Light and radio waves are types of radiation.

cosmic rays: radiation made up of high-speed, charged particles that come from outer space and hit Earth's atmosphere

combination V-2 and WAC Corporal, itself a smaller test-version of Corporal. On February 24, 1949, the two-stage rocket—the world's first—shot to a record altitude of 250 miles. Some historians have labeled this flight "the opening of the space age."

Caltech tapped Pickering to lead JPL in 1954. Even before he assumed the top post, the lab's engineers were dreaming of satellites. The year Pickering took over, JPL joined with von Braun's

Three key people in the history of the Jet Propulsion Lab, left to right, William Pickering, Theodore von Kármán, and Frank J. Malina

team to propose launching a satellite using von Braun's Redstone missile as the first stage and several JPL Sergeant missiles in the upper stages. "Project Orbiter" remained grounded, however; President Eisenhower wanted a nonmilitary space program, and both Redstone and Sergeant were weapons.

Eisenhower changed his mind when the Soviets orbited *Sputnik 1* in October and *Sputnik 2* in November 1957. On November 8, Pickering and von Braun were told to launch their satellite. They complied the following January, when their big rocket boosted *Explorer 1* to orbit. Two more Explorers followed over the next half-year. They confirmed what the first had discovered, that giant belts of radiation circled Earth. Scientists called these "Van Allen Belts" after James Van Allen, the physicist whose Geiger counters aboard the satellites took the historic readings.

Pickering and the lab weren't satisfied with their sudden fame. The United States was still losing the space race. On January 2, 1959, *Luna 1* launched by the Soviet Union became the first probe to escape Earth's gravitational pull, then the first to fly past the Moon. In October, *Luna 3* took the first photographs of the Moon's far side. But it was the Luna in between that rankled Pickering the most. On September 14, 1959, *Luna 2* actually hit the Moon.

The restless director knew JPL could do better. He wanted to explore the planets with robot spacecraft that could fly where they were told and return meaningful data on the way. The Lunas had done neither. None had had the ability to maneuver in space—they hit their targets because they were aimed at the right spot, not because anyone on the ground had told them where to fly. And Pickering did not consider *Luna 3*'s fuzzy pictures, however spectacular an engineering feat, to be meaningful to science.

The three men most responsible for the success of Explorer 1 *celebrate its January 1958 launch. From left are William Pickering, James Van Allen, and Wernher von Braun.*

Pickering faced an obstacle in building a planetary probe, however. In its race with the Russians, NASA had caught lunar fever. JPL was welcome to launch a Moon probe, Pickering was told, but a planetary machine would have to wait. That's when Pickering and his people had a clever thought.

NASA had assigned Project Ranger to JPL in December 1959. In a series of flights to begin in 1961, Ranger probes were to gather scientific information and snap pictures of the Moon as they plunged to the surface. JPL's idea was to design Ranger to be no simple Moon probe, but a prototype for a true planetary explorer.

Unfortunately, greater complexity meant greater risks. The first two Rangers, launched in August and November 1961, were stranded in Earth orbit by misfiring booster rockets. That increased the pressure for *Ranger 3* and *Ranger 4* to succeed. Part of the sense of urgency was funding—NASA could not long afford a failing project. An even greater concern, however, was extraterrestrial—Venus approached close enough to Earth to make exploration practical once every 19 months, and Mars every 26. The next launch window to Venus opened in August 1962 and the one to Mars a few months later. If JPL hoped to have modified Rangers ready to make the windows, the lab had to have some idea of what worked with their design and what didn't. Two failures would teach them little, and there would be no time to wait for *Ranger 5*. The Russians, meanwhile, would have Venus and Mars—and the history books—all to themselves.

booster rocket: a rocket stage that powers a spacecraft during liftoff, jettisoned (left behind) after the fuel is gone

trajectory: the path or curve made by a planet or spacecraft as it moves through space

THE BREAKTHROUGH

Hitting the Moon a quarter-million miles away was tough stuff. To get close enough to the right flight path, or trajectory, a Ranger probe had to hit a certain point in space (an imaginary 10-mile-wide circle 120 miles up) at just the right speed—between 24,520 miles per hour and 24,540, depending on the date of launch. As long as the Ranger's booster pushed the probe to within 16 mph of the target speed, Ranger could make up the difference with its onboard rocket engine. If it was moving outside that range, either too fast or too slow, no correction would help; Ranger would sail by the Moon and go into permanent orbit around the Sun.

Luna 2 had proven that probes didn't always need a midcourse correction to hit the Moon. Since this wasn't likely to be the case during a much longer planetary journey of tens of millions of miles, Ranger had the onboard rocket. To stabilize itself so the rocket would fire in the right direction, Ranger was fitted with tiny gas jets. This was something new: JPL's Explorers had spun on their axes like tops for stability. When Ranger wasn't aligning itself for the course correction, it used its stabilization system to point solar panels at the sun to recharge its batteries—another innovation. Recharging wasn't necessary for a three-day Moon flight, but it would be on a trip to Venus or Mars that would take months. Finally, Ranger's upgraded stability allowed it to communicate through a dish antenna pointed Earthward—again, not a device necessarily needed

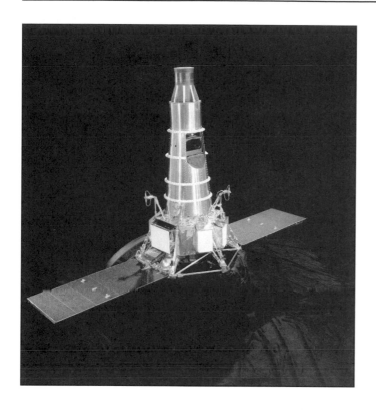

The Ranger probes provided over 17,000 images of the lunar surface, allowing scientists to study the Moon in great detail in preparation for the Apollo missions.

by a lunar probe, which could communicate just fine with a smaller multidirectional antenna.

Ranger 3 blasted off on January 26, 1962. Within minutes ground controllers knew the booster's first stage was pushing the probe too fast. *Ranger 3* would miss the Moon by about 20,000 miles. But all was not lost. No spacecraft had yet executed a midcourse correction. One would not help *Ranger 3* now—it was too far off course—but it could prove the probe's ability to perform the maneuver. Ground control radioed the necessary commands the next day and *Ranger 3* complied. Inexplicably, however, the probe moved in exactly the wrong direction. It was later determined that a programming error had caused the problem. Once they fixed the glitch, Pickering's people knew they had a spacecraft they could guide through deep space.

JPL also learned that no matter how good a spacecraft is, it will not perform correctly if its booster malfunctions. All the lab could do was prepare its next probe and hope for the best. *Ranger 4* lifted off from Cape Canaveral, Florida, on April 23. The first stage burned out and fell away, leaving the second stage to push the probe to orbit. *Ranger 4* coasted at 18,000 mph for several minutes until the second stage refired, boosting the probe to 24,520 mph. *Ranger 4* hit the bull's-eye of its 10-mile-wide target at precisely the right speed. Even without a midcourse correction, *Ranger 4* was on a path to become the first American spacecraft to hit the Moon.

And that's a good thing, because the probe's computer died on the way, disabling its onboard rocket. Still, when *Ranger 4* struck the Moon two days later, the press hailed its suicide flight as a major victory. "The fact that the Moon was reached," boasted *The New York Times*, "tells much about the increased power and improved accuracy the United States has achieved in rocket technology." True, the papers were starving for positive news from NASA, which had just orbited its first astronaut (John Glenn in February) more than 10 months after the Soviets. Despite their rosy outlook, the reporters were right that *Ranger 4*'s booster had proven its mettle. All the pieces were now in place for JPL to send its souped-up Ranger to Venus.

THE RESULT

The Soviets had no intention of letting Pickering and JPL beat them to Venus. In February 1961, they launched *Venera 1* and three months later, it became the first probe to fly by Venus, at a distance of 62,000 miles. Another demoralizing defeat for the United States, right? The Soviets thought so, but the fact is, *Venera 1*'s orientation system died 4.7 million miles from Earth, preventing the probe from turning its radio toward Earth. All communication to and from the spacecraft ceased. Rendered senseless, it zoomed past its target and into orbit around the Sun.

Venera's misfortune cleared the way to Venus for Pickering and JPL. The lab prepared a new Ranger for interplanetary flight, renaming it Mariner. Unfortunately, five minutes after *Mariner 1* blasted off in July 1962, it was blasted *apart*, destroyed by the range safety officer when its booster veered off course. Luckily, the launch window remained open for a month. A second robot explorer was hastily built and it rocketed skyward on August 27. For several tense minutes it seemed the rocket would again fail. The problem was corrected and *Mariner 2* slipped into its proper trajectory. From there, the probe coasted to Venus, extending its antennas and solar panels exactly as ordered. On December 14, 1962, the 442-pound *Mariner 2* sailed by Venus at a distance of less than 25,500 miles, close enough for its sensors to give humans their first detailed look at Venus.

Mariner 2 was the first in a string of stunning planetary successes for JPL. *Mariner 4*, launched in

The Venera 1 probe weighed 293 pounds and was equipped with two solar panels and a spike radio antenna that looks like an umbrella. The Soviets launched a total of 16 Venera probes between 1961 and 1983.

As a "flyby" mission, *Mariner 2* was never meant to land on Venus. Nevertheless, the probe drew near enough to discover that Venus's atmosphere consists of carbon dioxide and that it lacks a magnetic field. The probe also measured extreme surface pressures and temperatures. The latter registered a scorching 800° F, high enough to melt lead!

November 1964, acquired much scientific data during the first successful flyby of Mars in July 1965. *Mariner 6* and *7*, in a dual-spacecraft mission, briefly encountered the red planet in July and August 1969. *Mariner 9* revealed even more about Mars in November 1971, when it became the first spacecraft to orbit another planet. That mission was followed by the first-ever flyby of Mercury, accomplished on three different occasions by *Mariner 10* in 1974 and 1975. The last in the series, the tenth Mariner mission was the first spacecraft to use the now-familiar method of "gravity assist," whereby a spacecraft swings by one planet (in this case, Venus) to gain speed to reach another. In the process, *Mariner 10* became the first probe to visit two planets. It remains the only one to have reached Mercury.

Technicians install the solar arrays on one of the Mariner probes. These rectangular panels contained thousands of solar cells that helped power the spacecraft.

Then there were the two Viking missions to Mars, NASA's first concentrated effort to find life on the red planet. Each Viking was a two-part spacecraft, made up of lander and orbiter. JPL engineers, building on what they had learned from Ranger and Mariner, designed the orbiters. A private firm, Martin Marietta, designed and built the landers. In July 1976, the lander from *Viking 1* detached from its orbiter and flew down to the surface. *Viking 2* followed in September. Both landers and orbiters took thousands of brilliant color photographs. In addition, the landers performed soil experiments.

By the time *Viking 1* landed, JPL had a new director. Pickering had left the lab in April, a few months after his 65th birthday, Caltech's retirement age for administrators. He spent the next two years establishing and heading a research laboratory at the University of Petroleum and Minerals in Saudi Arabia. He retired to Pasadena in 1978.

JPL space probes continue to lead the world across space. In August and September 1977, *Voyager 1* and *2* were launched on a "grand tour" of four of the five outer planets—Jupiter, Saturn, Uranus, and Neptune. After visiting the first two, *Voyager 1* followed a trajectory that has taken it to the very edge of the solar system, more than 7.5 billion miles away. *Voyager 1* is the most distant human-made object in space. Its twin, which flew by and photographed all four planets, is almost 6 billion miles away. Both are expected to enter interstellar space sometime after 2010.

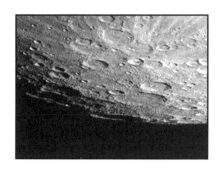

Mariner 10 captured this image of Mercury's south pole on its second flyby. The probe flew past the planet three times between March 1974 and March 1975, taking over 2,000 pictures.

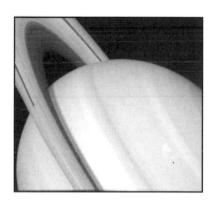

Voyager 2 returned this stunning view of Saturn and its rings in August 1981.

interstellar space: space among the stars, beyond our solar system

CHAPTER SIX

Sergei Korolev and the Soviet Space Program

The space race spawned brilliant innovations on both sides of the "Iron Curtain." In the United States, Wernher von Braun guided development of the Redstone rocket, then used it to boost the first U.S. satellite into orbit. William Pickering's people built the satellite, as well as America's first planetary probes. When astronauts rode von Braun's rockets into space, they traveled inside spacecraft designed under Max Faget.

The list of Soviet innovations is just as long. The list of Soviet innovators is not. One name— Sergei Pavlovich Korolev—stands out. Korolev was a codesigner of the Soviet Union's first liquid-fuel rocket. Later, he guided development of the world's first intercontinental ballistic missile (ICBM). Korolev's design bureau built the first artificial satellite, the capsule that carried the world's first space traveler, the first Soviet planetary probes, and the spacecraft that was to carry Soviet cosmonauts to

The moving force behind the Soviet space program, Sergei Pavlovich Korolev (1907-1966), was almost unknown during his lifetime. Referring to him only as the Chief Designer, the government kept his identity a state secret until after his death.

Laika ("barker" in Russian) rode Sputnik 2 *into space in November 1957. The little dog suffered no ill effects from the launch and weightlessness, but she died a week later when her air supply ran out.*

the Moon before the project was cut. And there's more. Korolev launched the first woman into space, the man who made the first space walk, the first reconnaissance and communications satellites, even the first dog. Indeed, it almost seems America ran the space race against one man.

In the first years of his life, which began in Zhitomir, Ukraine, on January 12, 1907, Sergei Korolev was not an especially happy child. His mother, Maria, had married Sergei's father, Pavel, by arrangement while she was still a teenager. The union failed before Sergei's third birthday and the boy was sent to live with Maria's parents. There were no other children for him to play with, and he spent long hours alone, building imaginary cities with blocks.

Sergei did enjoy school and learning. Even the turbulence of the Russian Revolution (1917) and the years of privation that followed could not stop him from devouring books on geometry, engineering, and literature. His enthusiasm intensified as he studied in one of the new Soviet vocational schools. It wasn't the instruction in laying tile that fascinated him, but the science he gleaned from his instructors, a mathematician and physicist among them.

By this time, Sergei's mother had remarried and with her husband and son had moved to Odessa, a Ukranian port city on the Black Sea. Stationed nearby was a detachment of military seaplanes. The craft were ragged at best, but Sergei fell in love anyway. He helped the mechanics and at times even flew with the pilots.

Korolev enrolled in the Kiev Polytechnical Institute to study aeronautical engineering (aircraft design) in 1924. He continued his education at the Moscow Higher Technical School from 1926 to 1929. His talent and devotion impressed Andrei Tupolev, an instructor and one of the Soviet Union's top aircraft designers. The day would come when Tupolev would prove his high regard for Korolev.

It was Korolev's passion for flight that led him to rockets. Likely he already knew about country-man Konstantin Tsiolkovsky, the famed rocket theorist. It was probably during this period that he learned of two other pioneers, Robert Goddard of the United States and Hermann Oberth of Germany. In 1930, Korolev joined the Group for Investigation of Reactive Motion (GIRD), an amateur group building a liquid-fuel engine. In August 1933, the organization, now led by Korolev, launched the Soviet Union's first liquid-fuel rocket.

The military quickly realized the rocket's destructive potential. It absorbed Korolev's group in October and renamed it the Reaction Propulsion Institute. Korolev was named second in command and chief designer, and was ordered to develop a pilot-less winged rocket. By late 1939, the Model 212 cruise missile—a rough equivalent of the German V-1 "buzz bomb" made famous later in World War II—had made several hundred test flights.

Korolev saw only the first of these. He was arrested in June 1938 on charges of sabotaging the institute's work. The charge was false, part of a titanic purge orchestrated by dictator Josef Stalin to

In a remarkable series of articles written between 1883 and 1935, Konstantin Tsiolkovsky (1857-1935) developed all the major principles that would one day govern space flight.

Sergei Korolev (farthest left, standing) and members of GIRD posed in 1933 with their GIRD-X rocket in a forest near Moscow.

terrorize and thus pacify his population. Korolev received a 10-year sentence at a labor camp in the Kolyma mountain range in northeast Siberia.

Korolev's prison, Kolyma, was part of the gulag, the system of work camps infamous for its murderous conditions. Indeed, Kolyma had a death rate of up to 30 percent a year, about 150,000 men.

Korolev and the others toiled in the camp's surface gold mine, cutting trees and digging in the ground. Malnutrition, exposure, disease, beatings, and executions were their lot. Korolev suffered a broken jaw, the loss of some teeth, and heart problems.

If not for Andrei Tupolev, Sergei Korolev likely would have vanished. After two years, Korolev was moved, at Tupolev's request, to a *sharashka*—a minimum-security prison whose inmates were provided with adequate food, shelter, and rest. This prison also housed Tupolev, himself a political inmate. Tupolev's sharashka teemed with scientists and engineers who went about their tasks designing aircraft. Over the next four years, Korolev designed planes for Tupolev and rocket engines and ballistic missiles in other sharashkas. He was a leading expert in rocketry by the time of his release in June 1944.

Andrei Tupolev, photographed in 1948

Although brilliant, Korolev and his comrades had developed nothing that compared in size and power to the German V-2, the world's first ballistic missile. From September 1944 through the following March, the Germans loosed 3,225 V-2s against Allied cities in England, France, and Belgium. Still, the device was disappointing as a weapon; on average, each took fewer than four lives. With improved range, accuracy, and explosiveness, however, advanced versions promised enormous destruction. Both the Americans and Russians understood this and scrambled to acquire V-2 technology after they defeated the Germans in May 1945. It was for this reason Sergei Korolev flew to Germany in September of that year.

As he picked through old blueprints and spare parts at the V-2 production facility, Korolev couldn't help thinking that had he not been arrested he might have equaled or surpassed the stunning success of the missile's creator, Wernher von Braun. From his first conversations with captured Germans, Korolev revealed an impatience to move beyond the V-2— and not just for use as a weapon. If we improve guidance, he told them, "and if we increase the range more and more, then we will finally be able to build artificial satellites which continue to orbit the Earth." Already he was thinking of a sputnik, but only as the first step into the cosmos: "If we increased the cut-off velocity [the rocket's ultimate speed] further by about 40 percent, then we could visit the Moon."

None of this surprised the Germans, who had heard the same from von Braun. What struck them as naive was Korolev's timetable: He seemed to want it all tomorrow. And he did. Korolev was in a race, but not with the enemies of communism. Since the terror of his imprisonment, he sometimes lay awake at night waiting for the banging of the dreaded security forces at his door. Korolev had seen death close up; he knew how suddenly it struck. *Time* was his enemy.

sputnik: the Russian word for satellite

THE BREAKTHROUGH

Upon Korolev's return from Germany in the summer of 1946, the military named him chief designer of long-range ballistic missiles. His first, the R-1, flew in October 1948. Other R-series rockets followed, including the R-7, the world's first ICBM.

Development of the Soviet ICBM is a perfect example of Korolev's ability to combine the good ideas of others into a better whole. When colleague Mikhail Tikhonravov suggested clustering engines for greater thrust, Korolev pounced on the innovation. Another peer, engine designer Valentin Glushko, liked the idea, but knew each additional engine meant another pump to feed fuel and oxidizer from the propellant tanks. Glushko, like all rocketeers, hated pumps. They were among the most problematic parts of an engine. His clever solution was to use a single, sturdy turbopump to feed several engines, thus slashing a missile's chances of failure.

As R-7 design was completed in 1953, Korolev figured the big missile would be able to boost a 3,000-pound satellite to orbit. From the beginning, however, his efforts to do that were resisted. A proposal to Communist Party leaders in 1953 mysteriously disappeared. A similar attempt in 1954 gained but a vague commitment to a "future" launch. The problem was, Soviet leaders weren't interested in space. They didn't seem to grasp that by orbiting the first satellite they could belittle the United States. Korolev did. In fact, it was central to his plans. He

intercontinental ballistic missile (ICBM): a long-range military rocket armed with a nuclear weapon, designed to make suborbital flights to give it enough range (over 4,000 miles) to target other continents. The Atlas and Titan rockets that launched the Mercury and Gemini spacecraft, respectively, were ICBMs adapted for space travel, as was Korolev's R-7 that sent *Sputnik* into orbit.

President Dwight David Eisenhower (1890-1969)

There was a legal precedent behind Eisenhower's "freedom of space." In international law, the principle of "freedom of the seas," first established by the Romans, forbids nations from claiming sovereignty, or control, of the ocean beyond their territorial waters, usually about 12 miles. Sometimes referred to as the "high seas," this area is free to all nations for naval and commercial navigation, and for fishing.

hoped that when a successful satellite launch brought the Soviets worldwide acclaim, the leadership would back even more spectacular space missions. Korolev would give his leaders the global prestige they desperately craved. In return, he would get the Moon and the planets.

Korolev's first big break came in July 1955 when the United States committed to launching a satellite during the upcoming International Geophysical Year (IGY). During the 18-month IGY, which would run from July 1957 through the end of 1958, scientists everywhere would try to learn more about Earth. America's motivation, however, had nothing to do with science. President Dwight Eisenhower feared a sneak attack by the Soviets and wanted spy satellites to warn of imminent danger. Overflying the Soviets with spacecraft, however, might well provoke them; according to international law, they owned the air over their nation. The law was less clear on whether or not they owned the space above the air. Eisenhower thought not and sought to establish the principle of "freedom of space." The U.S. would do this as nonaggressively as possible, by launching a civilian science satellite during the IGY. If the Soviets didn't erupt, Eisenhower would order up his spy satellites. Korolev's go-ahead to build his satellite was in response to the American effort.

As previously noted, the R-7 had a satellite "throw weight" (launch capability) of over 3,000 pounds. The Chief Designer intended to use every ounce. His "Object D" weighed 3,300 pounds, including 700 in scientific instruments. This was 15 times the weight of *Explorer 1*, the satellite the Americans were building.

Object D did not win the race to space; its complexity slowed its completion. Of course, *Explorer 1* didn't win, either. It was ready but its launcher wasn't. In August 1957, as the Americans stumbled with their satellite, Korolev initiated a last-minute replacement for Object D.

The two *prostreishiy sputniks*—simple satellites—his bureau built (one as a backup) were just under two feet in diameter and weighed 184 pounds. As work rapidly progressed on the devices, Korolev's missile engineers found themselves in uncharted waters. Their skills, tools, manufacturing facility, and testing procedures were all geared for making missiles, not satellites. This was no obstacle to Korolev, however. He demanded perfection in every task, no matter how small or seemingly irrelevant. Once, as engineers used a model sputnik to test the process of satellite separation from the missile, Korolev scolded the facility's chief for not polishing the mockup's surface. But this was not the actual sputnik, the brave worker countered. The Chief Designer, aware they were making history, shot back, "this ball will be exhibited in museums!"

Work on the sputniks ended in late September. On October 2, 1957, one of Korolev's big R-7s was rolled out of its assembly building at the Baikonur launch complex in the Soviet republic of Kazakhstan. Korolev walked in front of the missile as it crawled the nine-tenths of a mile to the launch pad. The R-7 was a "one-and-a-half-stage" rocket, with four

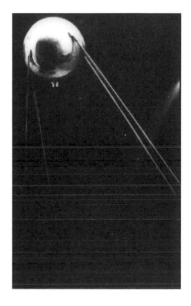

The ball that would be "exhibited in museums" and would jolt the world

engines in each of four strap-on boosters. Four additional engines were in the tail of the rocket itself. At a few minutes to midnight local time on October 4, all 20 engines roared to life. Two minutes later, as the missile climbed toward the northeast, the four boosters (the half-stage) burned out and dropped away. The remaining four engines burned for an additional two and a half minutes. As they, too, fell silent, the missile's nose separated, split apart, and exposed its spherical payload to open space. The first of the simple satellites was in orbit.

Sputnik, as it was called, had performed the first of two intended tasks—to be the first human-made object in space. Although Americans were unaware of *Sputnik*'s existence until it twice had passed over their country, they soon heard its scratchy beep-beep-beep loud and clear. Their response, reflected by their leaders, ranged from humiliation—"a devastating blow to the prestige of the United States as the leader of the scientific and technical world," said Henry L. Jackson, a leading senator—to fear as they realized that a missile that could launch a satellite around the world could drop an atomic bomb anywhere along the way.

Premier Nikita Khrushchev and his colleagues quickly realized that new space missions would further boost their empire's image of military and technical prowess. Just as Korolev had hoped, doors began to open before him. The Soviet space program became a reality. This had been *Sputnik*'s second mission. It, too, succeeded brilliantly.

Carrying Sputnik, *the Soviet R-7 rocket lifts off and ushers in the Space Age.*

THE RESULT

Khrushchev was so thrilled with *Sputnik* that he ordered another launch for November to coincide with the fortieth anniversary of the Communist revolution. *Sputnik 2* rocketed spaceward on November 3, 1957. At more than 1,100 pounds, it weighed six times the first. Part of the weight was a live dog, Earth's first orbiting space traveler. The dog soon died, but Khrushchev trumpeted the achievement to the world, nonetheless.

As long as Premier Khrushchev and colleagues remained happy, Korolev's funding increased. Over the next eight years, his bureau produced a staggering stream of "firsts"—first lunar probe (and the first to escape Earth's gravity), *Luna 1*, in January 1959; first spacecraft to carry a human, *Vostok 1*, in April 1961; first probe to fly by another planet (Venus), *Venera 1*, in May 1961; first spacecraft to carry a multi-person crew (three cosmonauts), *Voskhod 1*, in October 1964; and the first spacecraft equipped with an airlock (to allow a space walking crew member safe passage to and from the capsule), *Voskhod 2*, in March 1965.

Korolev selected space travelers, too, as chief of cosmonaut assignments. He personally chose Yuri Gagarin to make the first human space flight in April 1961; Valentina Tereshkova to become the first woman in orbit in June 1963; and Alexei Leonov to take humanity's first walk in space in March 1965. In every instance, variants of Korolev's R-7 provided the boost power.

Premier Khrushchev loved Korolev's "spectaculars." In addition to the Sputniks, Korolev's team also designed the Soviet Union's first spy satellite, Zenit, launched in April 1962, and its first communications satellite, Molniya, orbited in April 1965.

cosmonaut: an astronaut trained in the Soviet Union or, after 1991, Russia or the Republic of Kazakhstan; from the Greek words for "universe sailor"

Top: Sergei Korolev and the cosmonaut he chose to be first in space, Yuri Gagarin. Right: the Vostok capsule Gagarin rode in. Left: Alexei Leonov steps into the vastness of space.

The Chief Designer never realized his ultimate goal, to send cosmonauts to the Moon and beyond. By mid-1965, Korolev had drawn up plans to send two cosmonauts to lunar orbit, with one to land. Basically, the project was a cheaper, simpler version of the Apollo missions. It never happened. In December 1965, Korolev entered a clinic for a routine checkup. Doctors discovered a precancerous growth in his intestine and scheduled surgery. As Korolev drifted off under the anesthesia on the morning of January 14, both patient and surgeon (none other than the Soviet minister of health) expected a routine operation. Things went terribly wrong, instead. Korolev's cancer was worse than believed and old gulag injuries forced last-second changes in procedure. He survived the operation, but died suddenly of a heart attack 30 minutes later.

Sergei Korolev's death dashed any hopes for a lunar spectacular. Vasily Mishin, Korolev's deputy, became the new Chief Designer. Although he was an accomplished scientist, Mishin "lacked his predecessor's legendary genius for working the Soviet bureaucracy to get the resources he needed," said *The New York Times*, reporting Mishin's death in 2001.

"We could not compete with you Americans," recalled one technician. "There was confusion, disorganization. Too many 'chiefs' and not one boss. You can't get anywhere without a big man in charge." And the conquering giant of the Soviet space program was gone.

Being one of Korolev's cosmonauts was always dangerous. Konstantin Feoktistov, Vladimir Komarov, and Boris Yegorov (left to right), the first multi-person crew in October 1964, flew not in a spacecraft designed to handle three people, but in a hastily converted Vostok originally intended to carry one. The men were so cramped that they could not wear pressurized spacesuits. This risky practice lasted until June 1971, when an air leak during reentry suffocated the three cosmonauts aboard Soyuz 11.

CHAPTER SEVEN

Dale Reed
and the Lifting Body

Once the Americans and Soviets had mastered long-range ballistic missiles in the late 1950s, shooting astronauts and cosmonauts into orbit became routine. Getting them back was just as uncomplicated: They simply dropped through the atmosphere like warheads. When they got close enough to the surface—Americans over the ocean, Soviets over land—parachutes did the rest.

There was, in theory, a more elegant way to reach the ground. Spacecraft could have been designed to fly back, maneuvering through the atmosphere like airplanes and landing horizontally on runways. With such craft, there would have been no need to rescue American astronauts from splashdowns or for Soviet cosmonauts to endure even more severe "crashdowns." And because these spacecraft could have landed like airplanes, they could have been reused.

Engineer Robert Dale Reed (b. 1930) holds a model of the M2-F1 lifting-body aircraft he developed. The real M2-F1 is behind him. His work influenced the space shuttle design and may be used in future spacecraft.

Still, space travelers dropped from the sky like stones. The reason was the method's simplicity. During the 1960s, each superpower needed the quickest fix to trump the other. Disposable "ballistic" capsules were good enough to orbit crews, carry space walkers, and taxi astronauts to the Moon. Flyable spacecraft were anything but simple. Like capsules, they had to be stable at extreme altitudes and speeds, and resistant to the incinerating heat of reentry. Like airplanes, they had to be stable at low altitudes and speeds, and capable of landing. Unfortunately, what capsules were good at, planes were not, and vice versa. How to combine the best of both vehicles—to create a flyable, landable capsule—was a challenge too difficult to master under the decade's time constraints.

An Apollo capsule splashing down. This method of reentry required a navy ship to be nearby with a helicopter to retrieve the craft.

That doesn't mean the challenge went ignored. By the early 1960s, both NASA and the U.S. Air Force were conducting wind tunnel tests on models of "lifting bodies"—wingless aircraft that generated lift with the shape of their bodies alone. The data looked promising. Lifting bodies would handle well at high speeds, and their lack of wings would greatly enhance their ability to survive reentry. On the downside, their stability at lower speeds—and therefore their ability to land—remained uncertain.

As is usually the case in this type of story, someone stepped forward to build a real lifting body. If the thing worked, it would offer NASA an attractive alternative for a future spacecraft, perhaps even for the shuttle it someday hoped to build. If it didn't work . . . well, NASA could always go back to yet another ballistic configuration. The thought of astronauts riding in cannonballs for the indefinite future didn't sit well with aircraft designer Dale Reed. He built the first lifting body.

Robert Dale Reed was born in Ketchum, Idaho, on February 20, 1930. His father, John Reed, was a truck driver and would often bring his son along on jobs. A favorite destination for both was the train yard. The pair loved the steam engines; by the time he was nine or ten, Dale could name every one.

At age 12, Dale fell for another machine—the airplane. Famed glider pilot John Robinson was in windy Ketchum to attempt a world altitude record with his sailplane, *Zanonia*. Dale was there, too, helping Robinson where he could and spending hours watching the glider ride wind currents around the mountains. Robinson set two world records that summer. Dale Reed found a calling. Dale enlisted the help of his mother, Blanch, who taught woodshop at his school, in crafting model gliders and airplanes. At 16, Dale began taking flying lessons. Within days, he recalled, he was "totally hooked on aviation."

After high school, Reed studied engineering at the University of Idaho. Although he had been an average student in high school, classes on aircraft design inspired him to make almost perfect grades in college. Reed graduated in 1953 and headed south to Edwards Air Force Base in California, home of the National Advisory Committee on Aeronautics's High Speed Flight Research Center. The NACA, precursor to NASA, oversaw development of the nation's newest airplanes. Reed's job was to take aerodynamic readings from the surfaces of experimental aircraft like the X-15. Designed to test the stress of near-space on machine and pilot, the X-15 rocket plane ultimately reached a speed faster than 4,500 miles per hour and an altitude of over 67 miles.

Reed loved working with the sleek aircraft, but his vision of flight soon overtook even the high-speed X-15. The growth of the space program in the late 1950s excited him, despite NASA's decision in 1958 to rely on ballistic capsules for human space flight. Like most aircraft designers, Dale Reed saw ballistic capsules as little more than manned cannonballs. He didn't advocate conventional wings for spacecraft; they would have burned off during reentry. He did think it possible to shape a capsule so that even with no wings it could glide through the atmosphere and land like an airplane.

Reed did not originate the concept. In 1952, H. Julian "Harvey" Allen of Ames Aeronautical Laboratory of the old NACA had conceived of the "blunt-body" reentry shape that later inspired Max Faget, designer of the Mercury ballistic capsule. A colleague of Allen's, Alfred Eggers, later determined that Allen's cone-shaped vehicle could gain lift if its shape were modified. Specifically, Eggers wanted to cut it in half lengthwise and fly it flat side up. His final configuration, presented in 1958 and called the M-2, showed a bubble cockpit toward the nose and two stabilizer fins sticking up from either side at the rear. Eggers never built his odd-looking "lifting

The X-15 set the altitude record for winged aircraft—67 miles in 1963—and it was fast, reaching a top speed of 4,520 mph (Mach 6.7) in 1967. The X stands for experimental.

body"; the idea faded after NASA chose Faget's simpler non-lifting design for Mercury the same year.

Eggers's idea didn't die, however. Researchers at NASA's Langley Research Center in Virginia and in the U.S. Air Force had heard of it and were testing models of their own design. None had taken their concepts beyond the wind tunnel to actual construction. Many doubted whether lifting bodies could land without some sort of deployable wings. Some even thought they'd need jet engines.

Reed knew about Eggers's idea, too. He thought the M-2 could fly without wings and land without engines. Both wings and engines would add complexity. And both would take up space and weight best used in other areas, for example, to carry a larger crew or bigger payload. If he could prove Eggers's design worked, it would provide NASA with perhaps the most efficient spacecraft configuration possible. In early 1962, Reed went to work.

Alfred J. Eggers in the wind tunnel at the Ames Research Center in Mountain View, California

THE BREAKTHROUGH

Reed started small, launching paper gliders down the hallways of the Flight Research Center. Encouraged, he built a small wooden model. He dropped it from rooftops, pulled it by hand like a kite, even towed it aloft behind a larger radio-controlled model airplane. Everything told him a full-size lifting body would fly.

Reed couldn't develop the prototype alone. He needed funding and use of the center's facilities. That meant he needed the support of Paul Bickle, the center's director. Reed also needed other engineers and craftsmen to help design and build the lifting body, and a test pilot to fly it.

Enlisting helpers turned out to be no problem as word of Reed's unusual project spread. Dick Eldredge, an engineer, was first to join. He was an expert in aircraft design and knew all the center's welders, machinists, and other craftsmen. The duo recruited Milt Thompson, one of NASA's top test pilots. It was Thompson who suggested the group contact Eggers, then deputy director of NASA's Ames Research Center in California. Eggers's support would certainly help, and his wind tunnels would come in handy when the time came to test the lifting body's handling characteristics.

Armed with preliminary blueprints for the prototype and a top-notch pilot willing to fly it, Reed approached Bickle. The center director was a record-holding sailplane pilot, but he hedged on this strange glider until Reed showed a home movie he had made of the model lifting body in flight. Bickle relented. The team set to work building the real thing in September 1962.

Dubbed the M2-F1, the aircraft was constructed in two parts: an inner structure of tubular steel and a detachable outer shell. Eldredge took charge of building the inner structure and recruited several engineers and metalworkers to do the job. The outer shell posed more of a challenge. Because Reed would tow the lifting body into the air with a truck for its first tests, he had to keep the weight down. Normal aircraft aluminum would be too heavy. Bickle suggested Reed contact a local sailplane builder who could fabricate the outer shell from wood.

Skeleton and skin were mated in January 1963. Unfortunately, the completed M2-F1 weighed in at 1,000 pounds, about 400 more than Reed and Eldredge had planned. The problem was that none of the center's trucks could pull the thing fast enough to get it airborne. "Once again," Reed remembered, "a volunteer came along who had the know-how that we needed." One of his friends was a racing enthusiast who helped the team acquire a Pontiac Catalina convertible that they had souped up at a local garage. What they ended up with—besides what Reed quipped was "probably the first and only government-owned hot-rod convertible"—was a vehicle able to tow the M2-F1 at 110 miles per hour, fast enough for it to fly.

Now it was Milt Thompson's turn to show his stuff. The first towing took place in March 1963. It

The 1963 Pontiac capable of towing the M2-F1 (shown here) across the landing site at Edwards Air Force Base. The site is a huge dry lake about 44 miles square that Reed calls "one of the best natural landing sites on the planet."

was not encouraging. Several times Thompson tried lifting off, only to be forced back down when the lifting body bounced from side to side, threatening to flip over. The problem turned out to be his control system. The team made the adjustments and sent the prototype back to the runway. The new system worked. The Pontiac roared across the runway, its manned "kite" trailing at an altitude of 20 feet.

So Reed's team moved on to air tows. On August 16, 1963, the team hooked the M2-F1 to an Air Force C-47 transport plane. As the big airplane sped down the runway, Thompson pulled the lifting body's nose up prior to lifting off and . . . snap! The tow line separated, forcing the little prototype to a halt. It was determined that the lifting body's tow hook needed a quick adjustment.

Thompson considered his good luck as he waited. Had he been airborne when the tow line released, but still below about 200 feet, he would have been caught in the M2-F1's "dead-man zone." As Thompson explained, the lifting body's speed would drop rapidly once the tow line released. To avoid plummeting like a stone (and likely dying in the process), he had to "dump the nose over" in a dive to pick up speed, then "flare" (or level out) just before touching down. The trick was, he needed at least 200 feet to do it.

Once the tow hook was fixed (it took about half an hour), the C-47 again rumbled down the runway with Milt Thompson in tow. Over the next 20 minutes, the two craft ascended to 12,000 feet. Finally it was time for Thompson to release the tow-line. "Watching from the ground, it seemed that the M2-F1 literally fell out of the sky," recalled Reed. That was normal; the lifting body had lift, but not that much. It was designed to go in steep before flaring in the last seconds. Thompson made a slow U-turn to the left and then, after straightening out, performed a practice flare at 9,000 feet. So far, so good. He pushed the nose back down and maneuvered another 180 degrees to the left. With the runway straight ahead, Thompson dove below 200 feet, leveled out a final time, and touched the M2-F1 to the ground. The first flight for the wingless lifting body was history.

The M2-F1 being towed aloft for its first flight, which lasted less than two minutes from tow-line release to touchdown.

THE RESULT

Air Force test pilot Chuck Yeager was the third pilot to fly the M2-F1. He was so impressed that he encouraged the air force to join NASA in lifting-body research. Jerauld "Jerry" Gentry, another Air Force test pilot, made the last M2-F1 flight in August 1966. With almost 500 towed flights behind them, Reed's team had proven a vehicle with no wings or engines could maneuver and land. The M2-F1 program was, as Reed noted, "the key to unlocking the door to further lifting-body programs." The team followed with a heavier, all-metal lifting body, the M2-F2 (and the almost identical

Pilot Charles "Chuck" Elwood Yeager (b. 1923) became famous when he broke the sound barrier on October 14, 1947, while flying the Bell X-1 rocket aircraft at more than 670 mph. In 1957, he flew the Bell X-1A at more than two and a half times the speed of sound.

Test pilot Milt Thompson (on ladder) climbs into the cockpit of the M2-F2. This lifting body was launched from under the wing of a B-52 airplane.

M2-F3). Meanwhile, NASA's Langley Research Center built and tested its own version of the unpowered lifting body, the HL-10. The U.S. Air Force developed still another variant, the X-24A.

As he had hoped, Reed's initiative led to bigger and better things at NASA. The first was the space shuttle, which flew initially in 1981. Although not a true lifting body—it has wings, albeit thick ones angled back in a "delta" configuration to help them survive reentry—the shuttle does rely on the shape of its fuselage, or body, for lift. It's flyable from space to the ground, it glides through the air unpowered, and it makes pinpoint horizontal landings on a runway. All of these characteristics were proven possible by the lifting bodies.

Above: a full-scale model of the HL-10 mounted in a wind tunnel at Langley. Below: the space shuttle approaching the runway at Edwards Air Force Base.

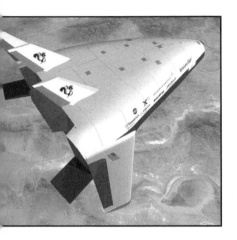

An artist's vision of what the X-33 or VentureStar spacecraft would have looked like had it been built.

When NASA announced a design for a replacement shuttle in 1996, it chose a true lifting body. The new orbiter, to be named VentureStar, was envisioned as the first "single-stage-to-orbit" spacecraft—that is, one able to jump to orbit under its own power, with no additional boosters. It would be shorter than the current shuttle (127 feet to 184 feet) and lighter (2.1 million pounds versus 4.5 million), yet still manage to haul about the same size payload to orbit (40,000-50,000 pounds). VentureStar's secret was in its lifting-body design: No wings meant more weight and space available for payload. This was expected to reduce payload launch costs from the shuttle's $5,000-$10,000 per pound to about $600.

Unfortunately, VentureStar proved too big a leap into the future. NASA, citing skyrocketing development costs, canceled the project in early 2001. The space agency may eventually choose another replacement. When it does, Reed is sure a lifting body is still the most logical choice; no shape can pack as much payload in as lightweight a package.

The lifting-body story does not end there. The X-24A shape developed by the air force was eventually chosen for the Crew Return Vehicle currently under development for the *Alpha* international space station. Should an emergency force crew members to abandon the outpost, they'll make their getaway in an X-24A-derived "lifeboat" called the X-38.

As for Reed, he became head of advanced planning at the Flight Research Center in 1964. Two vehicles he developed stand out. The first was a

lifting body called *Hyper 3*. In 1969, Thompson piloted its maiden flight, controlling it remotely from a cockpit set up on the ground. Thompson's radio-controlled flight proved that research vehicles could be flight-tested safely and relatively cheaply. The concept caught on as other prototypes were remotely tested. One was a 3/8 inch scale model of the F-15 *Eagle*, tested in 1974. The F-15 is now the U.S. Air Force's prime fighter jet.

Dale Reed and the X-38 Crew Return Vehicle he's helping to test

Yet another remote-control airplane was the *Mini-Sniffer*, built in 1975 to test the possible polluting effects of high-flying supersonic aircraft. Reed drew on his long history of building model airplanes and gliders when he created *Mini-Sniffer*. It weighs less than 190 pounds and has a 22-foot wingspan, both traits that give it maximum lift in Earth's thin upper atmosphere. And he didn't forget his first love, steam engines, when he needed to power the thing. *Mini-Sniffer*'s engine uses a super-hot gas (hydrazine) to operate a piston that in turn rotates a propeller. This is the same way a traditional steam engine works, but with hydrazine instead of water vapor. Most importantly, the hydrazine engine doesn't need oxygen.

The Mini-Sniffer III *crew wear self-contained suits because the hydrazine that powers the vehicle is extremely hazardous. This* Mini-Sniffer *flew to 20,000 feet.*

Put these characteristics together—the ability to operate in a thin atmosphere and the ability to operate with no oxygen—and you've got an aircraft that can fly through the atmosphere of Mars! By the late 1970s, other innovators had picked up on Reed's idea and designed different Marsplanes. Although NASA's drive to send one of the aircraft to Mars waned over the next decade or so, the idea is again . . . well, picking up steam. Within 10 years, offspring of the original *Mini-Sniffer* might well soar through Martian valleys, laden with sophisticated cameras and sensors.

Reed continues to work as an aerospace scientist at NASA's Flight Research Center, which was renamed the Hugh L. Dryden Flight Research Center in 1976. One project involves developing inflatable wings. An obstacle to Marsplane construction has been the aircraft's wings, which have to collapse so the airplane can fit inside a protective capsule during the fiery descent to the Martian surface. (All planets' atmospheres cause frictional heating, not just Earth's.) So far, solid wings that fold out have proved unworkable. Reed believes wings that inflate once the aircraft sheds its capsule could be the answer. He is working on other projects, too, such as aircraft that gather energy from ground-based lasers fired against their bellies. The laser energy would drive electric engines. Such environment-friendly propulsion could replace petroleum-derived fuels in the commercial airliners of tomorrow. Perhaps one day all of us will fly in aircraft with designs inspired by Dale Reed.

The Next Steps into Space

Where will the next great breakthroughs in space flight occur? They'll come in three areas: Earth-orbital, interplanetary (within our own solar system), and interstellar (beyond our system). Each realm will see unique advances.

Innovation within the first area, flight around our planet, will remain focused on the space station *Alpha* and its supporting fleet of reusable shuttles. Now in their third decade of operation, NASA's shuttles are old technology and approaching retirement. VentureStar, the next-generation shuttle cancelled in 2001 for lack of funding, would have made Earth-to-orbit transportation more routine by reaching orbit without expensive and cumbersome throwaway boosters. Perhaps much of VentureStar's breakthrough technology will find its way into the eventual shuttle replacement. Before that happens, however, a variant of the big lifting body may fly for the U.S. Air Force as the world's first space bomber.

Our present, above: the international space station Alpha. Funded and used by several nations, including the U.S., the station was transported in sections and then assembled in space. Our future, below: a rocket being launched by a magnetic levitation system. NASA researchers are currently testing this "maglev" method of using opposing magnetic polarities to lift a metal sled carrying a vehicle into space.

123

An armed version of VentureStar could fly around the world in 90 minutes, dropping precisely targeted bombs anywhere along the way from 60 miles above Earth.

Within a month of NASA's cancellation of the VentureStar project, military officials had expressed interest in acquiring the technology.

There are other breakthroughs in store for Earth-orbital flight. Currently, people and cargo (like satellites, telescopes, and supplies for *Alpha*) ride into space atop expendable boosters or inside the shuttle. It's a simple and effective method and, therefore, not likely to change anytime soon. Returning to Earth is another story. In the next few years, scientific payloads may drop from orbit by means of long tethers. As the package is lowered several miles below the space station, it will enter a lower orbit that will require an increase in speed to maintain. Without the needed boost, which it intentionally won't get, the package once released will drop from orbit to a preselected landing spot. Just like that—deorbit without having to lug along an extra engine and fuel. NASA successfully tested tether-assisted reentry in the early 1990s.

Tethers are versatile, too; besides sending objects home, they can boost them to higher orbits or even to the Moon and the planets. Imagine a tether cartwheeling in a low Earth orbit. After acquiring a payload from a shuttle, it slings the package into a higher orbit. There, a second tumbling tether captures the payload and whips it all the way to the Moon. Once in the grip of a third, lunar tether, the package can be sent to the surface or on to Mars and beyond. Tethers can even be dragged by spacecraft to generate electrical power for propulsion.

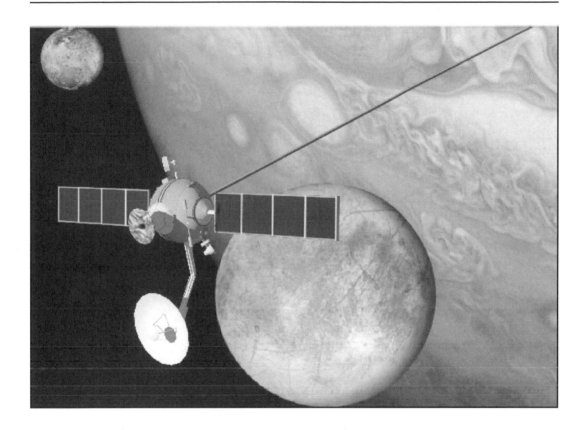

The next area for innovation is interplanetary space flight, where routine travel by humans likely will have to wait for a propulsion breakthrough like the Variable Specific Impulse Magnetoplasma Rocket. NASA is currently developing the technology. Plasma rockets don't burn chemicals to create thrust like today's rockets. Instead, gas like hydrogen is heated by electrical energy to temperatures from 50,000° F into the millions. The super-hot gas, called plasma, vents out the back of the rocket at a much greater velocity than that achieved by burning propellants. A magnetoplasma ship will travel faster and

An illustration of a spacecraft using a tether assist

Consider this analogy from a NASA website dedicated to interstellar travel: If the Sun were the size of a typical 1/2-inch diameter marble, the distance from the Sun to Earth, called an "Astronomical Unit (AU)" would be about 4 feet, Earth would be barely thicker than a sheet of paper, and the orbit of the Moon would be about 1/4 inch in diameter. On this scale, the closest neighboring star is about 210 miles away.

be able to maintain the speed far longer than a chemical rocket. The result: three to four months for a one-way trip to Mars, about six months less than currently possible.

Another possible engine for interplanetary travel would run on antimatter. It would take 1,000 external fuel tanks used by the current shuttle to contain the potential energy from just one gram—1/28 ounce!—of antimatter. A few billionths of a gram of the stuff could send a crew of astronauts to Mars in just six weeks or to Jupiter in less than eight months. Like magnetoplasma or chemical rockets, antimatter engines would be reactive devices. In one version, hydrogen gas heated by antimatter reactions jets rearward, pushing the craft forward; in another, antimatter explosions behind the ship shove it along its course. NASA is currently researching antimatter options in conjunction with universities and the military.

While magnetoplasma and antimatter rockets might propel us to the edge of our own star system in the next few decades, they won't be practical for interstellar travel. The star nearest to Earth is Proxima Centauri. It's 4.3 light-years away, about 24 trillion miles. A Saturn 5 could make the trip in 900,000 years. An antimatter rocket would need 900.

The propellant picture doesn't look rosy, either. The amount of fuel these rockets would consume would be prohibitive. In fact, it would be impossible to make enough propellant—there's not enough mass in the universe—to push a chemical rocket like the Saturn 5 to another star. As for antimatter, the necessary load would fill 10 railway tanker cars. And that's just to get to Proxima Centauri. If we want to slow down and explore, not to mention come back, we'll need a lot more antimatter.

We obviously need a major breakthrough in propulsion if we hope to visit just a few of the more than 200 billion stars in the Milky Way, our home galaxy. Since fuel is so big an obstacle, we'll need to find a way to propel a ship without it—a difficult challenge, but not impossible. A breakthrough in speed will be tougher to achieve. According to the laws of physics, nothing can move faster than the speed of light, about 186,000 miles per second. Even if we do attain this maximum speed, it will still take us almost four and a half years to reach the star next door. It seems as if we'll have to find a way to circumvent the laws of physics to travel faster than light.

NASA is currently working on these and other problems associated with interstellar space flight. For travel with little or no propellant and close to (but not faster than) the speed of light, scientists are considering various types of "sails" that would catch space radiation and use it as a propulsive force. In one example, the process would work like this: Radiation striking the spacecraft from the front is absorbed or neutralized, while the radiation colliding from the rear is reflected. The result is that the radiation from behind pushes the vessel forward. This is similar to how conventional sailboats take advantage of the wind.

To break through the light barrier, researchers are studying fantastic ideas straight out of last night's

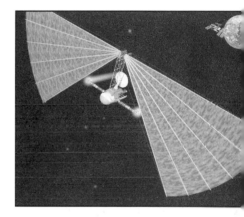

One concept of a spacecraft powered by electricity generated by solar cells on its large sails

rerun of *Star Trek*. If we could create them, wormholes, for example, might someday offer humans an instantaneous expressway to any point in the universe. First, we would have to bend space to move the point we wanted to reach nearer to us—like folding a sheet of paper. Then, with space folded into a giant "U", we would build our wormhole (really a worm "tunnel") from one leg to the other and fly our spaceship through. Of course, don't expect travel like this anytime soon. Scientists still don't know if it's possible to bend space or to erect wormholes.

Another intriguing idea is warp drive, Captain Kirk's preferred method of propulsion. How would it work? Since we can't move through space at a speed faster than light, the trick will be to wrap ourselves in a small bit of space and to accelerate that to beyond light speed. Confused? Consider climbing an escalator as you would a normal stationary staircase. You reach the top of an escalator faster because the stairs are moving with you. That's the idea behind warp drive. The problem is that scientists don't know if a "warp bubble" could be made to envelop a ship or if that patch of space could be made to move faster than light. They believe there was a time when space did move faster than light—as it exploded immediately after the "big bang," the hypothesized instant when the universe began its current expansion from a small cluster of very hot, very dense matter. The theory is, if it happened then, why not now? Again, these are obstacles that won't be overcome within the foreseeable future. But dreamers everywhere will continue to try.

Marc Millis, head of NASA's Breakthrough Propulsion Physics project at Glenn Research Center in Cleveland. His job is to invent a space drive that can carry travelers faster than the speed of light, something that most fellow rocket scientists think is impossible.

GLOSSARY

ablate: to burn or melt away

altitude: the height of something above a specific reference level

antimatter: according to theory, matter that is identical to physical matter except that its atoms are composed of antielectrons, antiprotons, and antineutrons ("anti" means displaying the opposite characteristics)

astronaut: a person specially trained to fly, navigate, or work aboard a spacecraft; from the Greek words for "star sailor"

astronomy: the science that studies the stars, planets, and all other celestial bodies

axis: a straight line about which an object rotates

ballistic missile: a rocket that flies under power to a predetermined altitude and location. The engine then cuts off and the missile falls freely to its target.

blunt body: a spacecraft with a flattened, or rounded, nose designed to slow the craft down and reduce heat during reentry into Earth's atmosphere

booster rocket: a rocket stage that powers a spacecraft during liftoff, jettisoned (left behind) after the fuel is gone

capsule: a pressurized module for carrying astronauts on space flights. Also called a **command module**.

centrifuge: an apparatus in which humans or animals are enclosed and which is revolved to simulate space flight acceleration

combustion chamber: the part of a rocket's propulsion system consisting of a chamber in which the fuel and oxidizer are combined and ignited. The resulting hot gases expand and escape through the nozzle to produce thrust.

command module (CM): the section of an Apollo spacecraft containing the crew's living area and the instruments and equipment they need. It measures about 11 feet high and 13 feet in diameter and has room for three astronauts. Also called a **capsule**.

cosmic rays: radiation made up of high-speed, charged particles that come from outer space and hit Earth's atmosphere

cosmonaut: an astronaut trained in the Soviet Union or, after 1991, Russia or the Republic of Kazakhstan; from the Greek words for "universe sailor"

dock: to join two or more spacecraft in space

Earth orbit rendezvous (EOR): a proposed method of landing on the Moon by combining several parts of a moonship in Earth orbit, then flying the whole ship to the Moon. Pronounced as letters E-O-R, not as a word.

feed system: a system that routes liquid fuels from tanks in a rocket into the combustion chamber

g-forces or g's: a unit used to measure the force of gravity. Earth's gravity is a g-force of one.

glider: a heavier-than-air aircraft that flies with no engine

gravity: the attraction of the mass of a celestial body, such as Earth, for other bodies. Gravity creates a sense of weight and causes objects to fall.

inertia: the tendency of a body to remain at rest or of a body moving to stay in motion

intercontinental ballistic missile (ICBM): a long-range military rocket armed with a nuclear weapon, designed to make suborbital flights to give it enough range (over 4,000 miles) to target other continents

interplanetary space: space among the planets, within the solar system and outside the atmosphere of any planet or the Sun

interstellar space: space among the stars, beyond our solar system

jettison: to cast or throw something off a ship

launch pad: a fireproof platform used as a base from which spacecraft are launched into space

lifting body: a wingless aircraft that generates lift with the shape of its body alone

liftoff: the moment a rocket starts traveling into the sky

liquid-fuel rocket: a rocket propelled by liquid chemical fuels and oxidizers

liquid hydrogen: the liquid form of hydrogen created by cooling the gas to extremely low temperatures. Hydrogen is the lightest, simplest element and burns easily.

liquid oxygen: the liquid form of oxygen created by cooling the gas to extremely low temperatures. Oxygen is one of the most abundant elements on Earth.

lunar: anything relating to the Moon

lunar module or lunar lander (LM): a spacecraft designed to carry astronauts from the Apollo command module to the Moon's surface and back

lunar orbit rendezvous (LOR): the method NASA chose to send astronauts to the Moon, requiring a three-part spacecraft to be launched from Earth to lunar orbit. Pronounced as letters L-O-R, not as a word. *See also* **command module** and **lunar module**

mockup: a model of a structure, usually full-sized and used for testing or demonstration

NACA: the National Advisory Committee for Aeronautics that was established by Congress in 1915 and became part of NASA in 1958. Pronounced as letters N-A-C-A, not as a word.

NASA: National Aeronautics and Space Administration, established October 1, 1958, to direct U.S. exploration of space. Pronounced as a word, NAS-sa, not as letters.

nozzle: the part of a rocket designed to vent the exploding exhaust gases from the combustion chamber rearward to cause forward motion

orbit: the path of one body as it circles another, such as Earth around the Sun, or a satellite around Earth

orbiter: the piloted section of the space shuttle that travels in space and lands like an airplane

oxidizer: an oxygen-rich substance that enables fuels to burn in space

payload bay: the area of the space shuttle used to carry cargo, located behind the crew compartment and in front of the engines. In orbit, double doors in the payload bay can be opened to deploy, or release, cargo.

probe: a spacecraft designed to leave Earth and explore and photograph planets in our solar system, their moons, and other objects in space

propellant: the combination of fuel and oxidizer that a rocket engine burns to achieve thrust

radiation: the giving off of energy in the form of rays or waves. Light and radio waves are types of radiation; so are streams of particles emitted by the atoms and molecules in a radioactive substance such as uranium.

rendezvous: the close approach of two spacecraft in orbit. The two craft stay close together and sometimes dock with each other.

retrorocket: a rocket engine that fires in the direction the spacecraft is traveling in order to slow the craft down. *See also* **rocket**

rocket: a reactive device that moves by burning fuel and expelling the resulting hot gases from one end. The rocket travels in the opposite direction from the escaping gases.

satellite: a small celestial body that orbits a larger one, such as the Moon traveling around Earth. An artificial satellite is a human-made object, such as a weather satellite, that orbits Earth.

service module: the section of an Apollo spacecraft that contained the spacecraft's main engine, fuel cells, water, and other supplies

space: the vast region beyond the atmosphere of Earth

space shuttle: a spacecraft made up of an orbiter, two solid rocket boosters, and an external fuel tank. The orbiter travels through space and returns to Earth, gliding to a landing.

space suit: a protective garment with life-support and communication systems that enable the wearer to survive in space

space walk: any activity by an astronaut or cosmonaut outside the spacecraft or on the surface of the Moon

sputnik: the Russian word for satellite

staging: the use of two or more rocket modules, or stages. Rocket stages boost spacecraft to higher altitudes or help them carry heavy cargo.

suborbital: a flight made by a rocket or spacecraft in which the object reaches space, but does not travel fast enough to attain orbit and falls back to Earth in a long curve

thrust: the driving force that pushes a rocket engine forward

trajectory: the path or curve made by a planet or spacecraft as it moves through space

velocity: the rate at which a body moves in a certain direction, expressed in measures of distance and time

BIBLIOGRAPHY

Books

Armstrong, Neil, Michael Collins, and Edwin E. Aldrin Jr. *First on the Moon.* Written with Gene Farmer and Dora Jane Hamblin. Boston: Little, Brown, 1970.

Baker, David. *The History of Manned Space Flight.* New York: Crown Publishers, 1981.

Bergaust, Erik. *Wernher von Braun.* Washington, D.C.: National Space Institute, 1976.

Bergreen, Laurence. *Voyage to Mars: NASA's Search for Life Beyond Earth.* New York: Riverhead Books, 2000.

Bilstein, Roger E. *Stages to Saturn.* The NASA History Series: NASA SP-4206. Washington, D.C.: NASA, 1980.

Braun, Wernher von. *The Mars Project.* Chicago: University of Illinois Press, 1991. (first printed 1952).

———. *Space Travel: A History* (update of *History of Rocketry and Space Travel*). With Frederick I. Ordway III and Dave Dooling. New York: Harper & Row, 1985.

———. *The Rockets Red Glare: An Illustrated History of Rocketry Through the Ages.* With Frederick I. Ordway III. New York: Anchor Press/Doubleday, 1976.

Breuer, William B. *Race to the Moon.* Westport, Conn.: Praeger Publishers, 1993.

Brooks, Courtney G., James M. Grimwood, and Loyd S. Swenson Jr. *Chariots for Apollo: A History of Manned Lunar Spacecraft.* The NASA History Series: NASA SP-4205. Washington, D.C.: NASA, 1979.

Brown, Kenneth A. *Inventors at Work.* Redmond, Wash.: Tempus Books of Microsoft Press, 1988.

Burrows, William E. *This New Ocean: The Story of the First Space Age.* New York: Random House, 1998.

Clark, Phillip. *The Soviet Manned Space Program.* New York: Orion Books, 1988.

Coil, Suzanne M. *Robert Hutchings Goddard: Pioneer of Rocketry and Space Flight.* New York: Facts On File, 1992.

Crouch, Tom D. *Aiming for the Stars: The Dreamers and Doers of the Space Age.* Washington, D.C.: Smithsonian Institution Press, 1999.

Damon, Thomas D. *Introduction to Space: The Science of Spaceflight.* Malabar, Fla.: Krieger Publishing, 1989.

Dethloff, Henry C. *Suddenly, Tomorrow Came . . . : A History of the Johnson Space Center.* The NASA History Series: NASA SP-4307. Houston: NASA, 1993.

Dunar, Andrew J., and Stephen P. Waring. *Power to Explore: A History of Marshall Space Flight Center 1960-1990.* The NASA History Series. Online at http://history.msfc.nasa.gov/book/bookcover.html.

Farley, Karin Clafford. *Robert H. Goddard.* Englewood Cliffs, N.J.: Silver Burdett Press, 1991.

Goldstein, Stanley H. *Reaching for the Stars: The Story of Astronaut Training and the Lunar Landing.* New York: Praeger Publishers, 1987.

Hacker, Barton C., and Charles C. Alexander. *On the Shoulders of Titans: A History of Project Gemini.* The NASA History Series: NASA SP-4203. Washington, D.C.: NASA, 1977.

Hall, R. Cargill. *Lunar Impact: A History of Project Ranger.* The NASA History Series: NASA SP-4210. Washington, D.C.: NASA, 1977.

Hansen, James R. *Spaceflight Revolution: NASA Langley Research Center from Sputnik to Apollo.* The NASA History Series: NASA SP-4308. Washington, D.C.: NASA, 1995.

Harford, James. *Korolev: How One Man Masterminded the Soviet Drive to Beat America to the Moon.* New York: John Wiley, 1997.

Heppenheimer, T. A. *Countdown: A History of Space Flight.* New York: John Wiley, 1997.

Jenkins, Dennis R. *Space Shuttle: The History of Developing the National Space Transportation System.* Cape Canaveral, Fla.: Dennis R. Jenkins, 1997.

Kluger, Jeffrey. *Journey Beyond Selene: Remarkable Expeditions Past Our Moon and to the Ends of the Solar System.* New York: Simon & Schuster, 1999.

Koppes, Clayton R. *JPL and the American Space Program.* New Haven, Conn.: Yale University Press, 1982.

Launius, Roger D. *Frontiers of Space Exploration.* Westport, Conn.: Greenwood Press, 1998.

Lehman, Milton. *This High Man: The Life of Robert H. Goddard.* New York: Farrar, Straus, 1963.

Lewis, Richard S. *The Voyages of Columbia: The First True Spaceship.* New York: Columbia University Press, 1984.

Logsdon, John M. *The Decision to Go to the Moon.* Cambridge, Mass.: MIT Press, 1970.

Mauldin, John H. *Prospects for Interstellar Travel.* Vol. 80, Science and Technology Series. San Diego: American Astronautical Society, 1992.

Murray, Charles, and Catherine Bly Cox. *Apollo: The Race to the Moon.* New York: Simon & Schuster, 1989.

Neufeld, Michael J. *The Rocket and the Reich: Peenemünde and the Coming of the Ballistic Missile Era.* Cambridge, Mass.: Harvard University Press, 1995.

Newkirk, Dennis. *Almanac of Soviet Manned Space Flight.* Houston: Gulf Publishing, 1990.

Nicks, Oran W. *Far Travelers: The Exploring Machines.* NASA SP-480. Washington, D.C.: NASA, 1985.

Oberg, James E. *Red Star in Orbit.* New York: Random House, 1981.

———. *The New Race For Space.* Harrisburg, Penn.: Stackpole Books, 1984.

———. *Uncovering Soviet Disasters.* New York: Random House, 1988.

Ordway, Frederick I., III, and Mitchell R. Sharpe. *The Rocket Team: From the V-2 to the Saturn Moon Rocket.* Cambridge, Mass.: MIT Press, 1979.

Pellegrino, Charles R., and Joshua Stoff. *Chariots for Apollo: The Making of the Lunar Module.* New York: Atheneum, 1985.

Piszkiewicz, Dennis. *Wernher von Braun: The Man Who Sold the Moon.* Westport, Conn.: Praeger Publishers, 1998.

Reed, R. Dale. *Wingless Flight: The Lifting Body Story.* NASA SP-4220. Washington, D.C.: NASA, 1998. Online at http://www.dfrc.nasa.gov/History/Publications/WinglessFlight/.

Schefter, James. *The Race.* New York: Doubleday, 1999.

Segel, Thomas. *Men In Space.* Boulder, Colo.: Paladin Press, 1975.

Shepard, Alan, and Deke Slayton. *Moon Shot: The Inside Story of America's Race to the Moon.* Atlanta: Turner Publishing, 1994.

Smith, Arthur. *Planetary Exploration.* Wellingborough, England: Patrick Stephens Limited, 1988.

Stuhlinger, Ernst, and Frederick I. Ordway III. *Wernher von Braun, Crusader for Space: A Biographical Memoir.* Malabar, Fla.: Krieger Publishing, 1994.

————. *Wernher von Braun, Crusader for Space: An Illustrated Memoir.* Malabar, Fla.: Krieger Publishing, 1994.

Swenson, Loyd S., Jr., James M. Grimwood, and Charles C. Alexander. *This New Ocean: A History of Project Mercury.* The NASA History Series: NASA SP-4201. Washington, D.C.: NASA, 1989.

Thompson, Milton O., and Curtis Peebles. *Flying Without Wings: NASA Lifting Bodies and the Birth of the Space Shuttle.* Washington, D.C.: Smithsonian Institution Press, 1999.

Winter, Frank H. *Prelude to the Space Age—The Rocket Societies: 1924-1940.* Washington, D.C.: Smithsonian Institution Press, 1983.

————. *Rockets into Space.* Frontiers of Space series. Cambridge, Mass.: Harvard University Press, 1990.

Magazine and Newspaper Articles

Braun, Wernher von. "Crossing the Last Frontier." *Collier's*, March 22, 1952.

————. "Man on the Moon: The Journey." *Collier's*, October 18, 1952.

————. "Man on the Moon: The Exploration." Written with Fred L. Whipple. *Collier's*, October 25, 1952.

————. "Baby Space Station." Written with Cornelius Ryan. *Collier's*, June 27, 1953.

————. "Can We Get to Mars?" *Collier's*, April 30, 1954.

Cooper, Henry S. F., Jr. "Annals of Space: Max Faget and Caldwell Johnson." *The New Yorker*, September 2, 1991.

DiChristina, Mariette. "Space at Warp Speed." *Popular Science*, May 2001.

Douglass, Steve. "B-3 and Beyond." *Popular Science*, February 2000.

Hallion, Richard P. "The Space Shuttle's Family Tree." *Air & Space*, April/May 1991.

Houbolt, John C. "Lunar Rendezvous." *International Science and Technology*, February 1963.

Larson, George C. "The Next Generation." *Air & Space*, January 2000.

Leary, Warren E. "NASA May Recast Space Station to Cut Costs." *The New York Times*, online version, April 5, 2001.

Morton, Oliver. "MarsAir." *Air & Space*, January 2000.

Reinert, Patty. "Lawmakers Warn of Space Station Inertia." *Houston Chronicle*, April 5, 2001.

Schneider, Greg, and Kathy Sawyer. "New Mission for Lockheed Space Plane?" *Washington Post*, online version, April 13, 2001.

Sweetman, Bill. "VentureStar: 21st Century Space Shuttle." *Popular Science*, October 1996.

Wilkinson, Stephan. "The Legacy of the Lifting Body." *Air & Space*, April/May 1991.

Video Documentaries

The Idea That Nobody Wanted. NASA Langley Research Center: Video Applications Group.

To The Moon. (Public Broadcasting System/NOVA) NASA Langley Research Center: Video Applications Group.

Interviews

Faget, Maxime A. Interview with Robbie Davis-Floyd and Kevin Cox, September 10, 1996, for Institute for Advanced Interdisciplinary Research. Online at http://space.systems.org/oh/allthree/2_part_2.htm.

Faget, Maxime A. Interview with author, January 20, 2000.

Houbolt, John C. Interview with author, January 12, 2000.

Pickering, William H. Interview with Mary Terrall, November-December 1978. California Institute of Technology Oral History Project. Pasadena: California Institute of Technology, 1981.

Reed, Robert Dale. Interview with author, March 27, 2001.

INDEX

ABOUT THE AUTHOR

Jason Richie visited the Johnson Space Center in his hometown of Houston, Texas, to research this book as well as another Oliver Press title, *Spectacular Space Travelers.* That's a Saturn 5 rocket behind him. A former noncommissioned officer in the U.S. Army, Richie graduated *summa cum laude* from the University of Minnesota with a degree in American history. He lives in Houston with his wife, Diana, and son, James.

PHOTO ACKNOWLEDGMENTS

Clark University Archives: pp. 13, 18, 29
John C. Houbolt: p. 64
John F. Kennedy Library: p. 42
Library of Congress: pp. 9 (both), 10 (top), 20, 36, 60, 97, 100, 103
National Aeronautics and Space Administration (NASA): cover, pp. 16, 46, 47, 53 (both), 58, 74, 122 (top)
NASA Ames Research Center: pp. 55, 111
NASA Dryden Flight Research Center: pp. 106, 110, 114, 115, 116, 117 (bottom), 119, 120
NASA Glenn Research Center: pp. 57, 128
NASA Jet Propulsion Laboratory: pp. 78, 82, 84, 87, 90, 91 (both)
NASA Johnson Space Center: pp. 50, 62, 76 (both), 77, 108
NASA Kennedy Space Center: pp. 72, 80
NASA Langley Research Center: pp. 59 (top), 69, 117 (top)

NASA Marshall Space Flight Center: pp. 2, 15, 40, 44, 59 (bottom), 61, 71, 118, 122 (bottom), 125, 127, back cover
National Air and Space Museum (NASM), Smithsonian Institution: pp. 10 (bottom, SI Neg. No. NAS-4086E), 11 (76-17287, courtesy of RIA-Novosti), 12 (75-11483), 21 (77-6027), 25 (73-1274), 30 (A-1074), 31 (A-4968), 32 (A-4555), 33 (84-14740), 34 (76-13637), 38 (76-7559), 39 (86-13268), 49 (91-19620), 89 (NASM-A391D), 92 (76-17276, courtesy of RIA-Novosti), 94 (75-10226), 95 (A-4110-A), 96 (73-7133, courtesy of The Konstin E. Tsiolkovsky State Museum of the History of Cosmonautics, Russia), 104 (all) (top, 73-379, courtesy of RIA-Novosti), (left, 94-7709), (right, 74-12209, courtesy of RIA-Novosti)
Sovfoto/Eastfoto: pp. 6, 102, 105
TASS: p. 101